The Rise of Chance in Evolutionary Theory

The Rise of Chance in Evolutionary Theory
A Pompous Parade of Arithmetic

Charles H. Pence

Université catholique de Louvain

ELSEVIER

ACADEMIC PRESS
An imprint of Elsevier

Academic Press is an imprint of Elsevier
125 London Wall, London EC2Y 5AS, United Kingdom
525 B Street, Suite 1650, San Diego, CA 92101, United States
50 Hampshire Street, 5th Floor, Cambridge, MA 02139, United States
The Boulevard, Langford Lane, Kidlington, Oxford OX5 1GB, United Kingdom

Notices
Knowledge and best practice in this field are constantly changing. As new research and experience
broaden our understanding, changes in research methods, professional practices, or medical treatment
may become necessary.

Practitioners and researchers must always rely on their own experience and knowledge in evaluating
and using any information, methods, compounds, or experiments described herein. In using such
information or methods they should be mindful of their own safety and the safety of others, including
parties for whom they have a professional responsibility.

To the fullest extent of the law, neither the Publisher nor the authors, contributors, or editors, assume
any liability for any injury and/or damage to persons or property as a matter of products liability,
negligence or otherwise, or from any use or operation of any methods, products, instructions, or ideas
contained in the material herein.

Library of Congress Cataloging-in-Publication Data
A catalog record for this book is available from the Library of Congress

British Library Cataloguing-in-Publication Data
A catalogue record for this book is available from the British Library

ISBN 978-0-323-91291-4

For information on all Academic Press publications
visit our website at https://www.elsevier.com/books-and-journals

Publisher: Nikki Levy
Acquisitions Editor: Anna Valutkevich
Editorial Project Manager: Veronica III Santos
Production Project Manager: Kiruthika Govindaraju
Cover Designer: Greg Harris

Typeset by STRAIVE, India

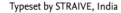

Working together
to grow libraries in
developing countries

www.elsevier.com • www.bookaid.org

Contents

Acknowledgments

To Phil Sloan and Jon Hodge, without whom I can assert, with probability one, that this book would never have existed

The work that would eventually become this book has been percolating steadily since I first jotted down some initial sketches on Karl Pearson and W. F. R. Weldon in 2008, and so I will be entirely unable to adequately pay my intellectual debts here. Phil Sloan has been a tireless friend and interlocutor on all of these matters for more than a decade. Greg Radick has been incredibly generous with his time and his firm belief that there is a field of Weldoniana to be built, if only we can construct it. (Yafeng Shan is also a crucial fellow traveler here.) Jon Hodge and Michael Ruse, in their quite different ways, have each been indefatigable in their support for a junior scholar's entry into the crowded field of the history and philosophy of biology. In the immediate process of writing and publishing, I should thank a number of audiences, including at the History of the Philosophy of Science (HOPOS) 2018 meeting in Groningen, the European Philosophy of Science Association (EPSA) 2019 meeting in Geneva, the New Directions in the Historiography of Genetics Conference at the Cohn Institute in Tel Aviv, the Aarhus Center for Science Studies, the PhilInBioMed group at the Université de Bordeaux, the McMaster University Philosophy Colloquium, and two local groups—the UCLouvain CEFISES Work in Progress Seminar and the LIBST and ELI-B groups in biology—all of which offered a variety of penetrating and exceptionally useful comments. Thony Christie offered excellent corrections to my poor German translations of Johannsen. Nils Roll-Hansen convinced me to look back over Yule's mathematics, for which I'm extremely grateful, as it led to a correction of a prior reading of Yule in the literature. Alex Aylward greatly assisted with my reading of Fisher, especially via his encyclopedic knowledge of the *Genetical Theory of Natural Selection*.

Work on this book was supported by a US National Science Foundation grant (NSF-SES #1826784). My deep and abiding thanks to what remains of American governmental and taxpayers' commitments to the life of the mind. I also thanks my colleagues at LSU and UCLouvain for their supererogatory efforts in setting up my work arrangements in the early days of the writing process.

Finally, and too briefly, my thanks are due to my family: to my parents, who form the best moral and editing support team that a guy could want, and to Julia, who confirms daily that I can (at least occasionally) choose well. Darwin wasn't wrong when he called marriage "the best & almost only chance for what share of happiness this world affords."

Chance governs the descent of a farthing: Charles Darwin

If I am as muzzy on all subjects as I am on proportions & chance,—what a
Book I shall produce!
Darwin, letter to John Lubbock, 14 July 1857

It is April of 1901, 19 years after the death of Charles Darwin, and Karl Pearson cautiously writes to Francis Galton. Along with W. F. R. Weldon and Charles Davenport, he has recently begun to hatch plans to launch the first journal entirely dedicated to the statistical study of biology, *Biometrika*, and he hopes for Galton's help—perhaps a position on the editorial board, an article for the first number of the journal as a "send off," even a contribution toward the guarantee fund (Pearson, 1901a)? If the journal, he says, can "survive the risk of infantile mortality, we will live on" (Pearson, 1901b). A few months later, with Galton's help (and his financial backing) secured, Pearson turns to the matter of situating the statistical study of evolution within its history. "I have been looking at one or two of Darwin's books," he writes to Galton, "to see if he anywhere emphasises the value of statistical enquiry. I can find nothing, and yet I feel quite certain he realised its value by undertaking, as he did, the long series of experiments in cross- and self-fertilisation of plants." Surely, somewhere, Pearson pleads, there must be an "apt remark as to the need of statistical method in solving evolution problems" (Pearson, 1901c)?

Galton has no good news. He writes that he well remembers discussing the matter of statistics with Darwin, but "I doubt if he ever thought very much or depended much on statistical inquiry in his own work" (Galton, 1901a). Before Pearson can respond, he adds that, having spoken to Darwin's children Francis and Leonard, "their views about their father's attitude toward statistics are the same as mine, except that Frank's was more strongly expressed. I fear you must take it as a fact that Darwin had no liking for statistics. They even thought he had a 'non-statistical' mind, rather than a statistical one" (Galton, 1901b). Pearson replies, clearly crestfallen. "Many thanks for your inquiries," he laments, "which I fear means we can find no statement of a definite kind." Darwin's view of statistics, Pearson muses through rose-colored glasses, must amount to a general realization "that the root of evolution lay in the *largeness of the numbers* dealt with" (Pearson, 1901d).

While we may no longer be able to write to Francis Darwin for confirmation, Pearson has the right idea—any study of the role of chance and statistics in evolutionary theory must begin with Darwin himself. As Pearson correctly sees, Darwin

The Rise of Chance in Evolutionary Theory. https://doi.org/10.1016/B978-0-323-91291-4.00007-8

was not, and could not have been, entirely ignorant of the role of chance and the utility of something like statistical inference in his development and subsequent elaboration of evolution by natural selection. But this leaves much room for interpretation. Indeed, one of the most striking differences between the evolutionary theory that we read in Darwin's work and the evolutionary theory of a 21st century textbook is a shift in the importance of statistical methodology and the roles for chance that are implied by its use. Darwin introduces natural selection in a rich context alongside a whole host of other topics, ranging from biogeography to embryology to paleontology, over the course of some 490 pages. And he does so with barely any reference to mathematics whatsoever (for an extensive survey of these few uses, see Sheynin, 1980). A contemporary textbook on evolutionary theory, on the other hand, makes it only three pages without a graph, and is describing mathematical models for phylogenetic inference by page 28 (Futuyma, 2005).

Understanding this shift is precisely the goal of this book. How did we move from an essentially non-statistical, non-mathematical theory of evolution, with only a few circumscribed roles for chance, to a thoroughly mathematized and statistical theory in which the interpretation of chance stands as one of the most significant philosophical issues? And what can this history tell us about evolution's present and future?

But let's begin at the beginning.

Darwin before the *Origin*

Scholars studying Darwin have, perhaps more than in the case of any other major figure in the history of science, a vast wealth of correspondence, notebook, draft, and manuscript materials on which to draw. We can thus trace, in minute detail, the development and maturation of Darwin's idea of evolution. At the very beginning of Darwin's career, as he sets off to make a name for himself as the "ship's naturalist" aboard HMS *Beagle* (in fact, little more than a rich, intellectual companion for the captain, whose predecessor had committed suicide at sea), he is a fairly typical, young English scientist. His interests focus on geology and natural history, and perhaps his most significant influence on both scores is Charles Lyell. Lyell's picture of the natural world gives pride of place to the harmony evidenced by the economy of nature, prodded by the occasional, yet ordered and regular, creation of new species. Darwin's first work on species may thus be profitably read as a series of breaks with his friend and mentor, as he abandons the tenets of Lyell to craft his own view, particularly in his notebook writings of 1837 and 1838.

This much is a fairly well known story, and to detail all of its contours (along with Darwin's debts to other authors in this era, especially German Romantics like Alexander von Humboldt) would be the project of several books (the interested reader may consult Ospovat, 1981; Hodge, 1983; Sloan, 1986; or Hodge, 2009, among many others). What is less well understood is the early development of Darwin's views of chance prior to the publication of his *opus magnum*. As we survey this period with chance as our focus, what we will find are fewer such breaks than he makes with Lyell on most other subjects. Lyell's own view—as should be clear from its emphasis on natural law, balance, and harmony—leaves precious little room for the working of chance in nature. While Darwin refashions his understanding of the creation, dispersal, modification, and relations of species over these years (each of which constitutes no small feat, much less all of them at once), chance remains for him an issue he does not quite know how to handle. The tension that emerges in this early period, constituting as it does a sort of unwelcome, doubtful guest in Darwin's theorizing, continues to loom large in his thought for the rest of his life.

As Darwin departs on the *Beagle* at the end of 1831, he carries with him the first volume of Lyell's *Principles of Geology* (inscribed as a gift from the ship's captain), the other volumes to arrive via post in the next few years of the voyage. Twenty-four months later, and shortly after the third volume was sent, Darwin's Lyellian bent is so well known that his sister Catherine writes to him that "I hear that your Theory of the Earth is supposed to be the same as what is contained in Lyell's 3$^\mathrm{d}$ Vol." (Darwin, 1833). What, then, is Lyell's view, which forms Darwin's point of departure for all of his initial thoughts on the origin and extinction of species? To begin, we must understand the focal "species problem" for Lyell. Lyell is a deeply committed uniformitarian—he argues, that is, that all natural phenomena that are now visible on the earth are the result of the same set of causes that still operate today (floods, volcanoes, earthquakes, and the like), acting at effectively the same intensity as they always have, back through an immeasurably vast geological past. For prior uniformitarians such as Hutton, this sufficed: the earth is a static and perfectly balanced creation from time immemorial, exemplifying the best features of divine providence. But by Lyell's day, any such uniformitarianism was confronted with the problem of extinction. While life on earth has broadly looked the same as it does now (with many major groups appearing to have been similar back to the deepest fossil layers known at the time), the species which constitute it have not. The earth has changed enough to render it uninhabitable for some forms—think woolly mammoths wandering the forests of North America—but not so much as to violate the broader precepts of uniformitarianism.

Lyell therefore needs to introduce just the right amount of variability into the natural order, and he does so by pointing to one major cause of species destruction, and a second major, compensating cause of species creation. The destroyer is the re-distribution of climatic patterns, along with biological competition. There is no con-tradiction with a general uniformitarian view if the pattern of climate, while retaining essentially the same overall features across geologic time, has modified its distribu-tion across the globe. New land will emerge, old land erode and subside, and with these developments (all observable as well, for still operating in the present) will undoubtedly come climatic differences. These changes will result, in turn, in altera-tions to inter-species relations. Species will go extinct, as the finely tuned conditions necessary for their existence disappear (or move outside their range) and they fall victim to competition. To retain long-term balance, we must offset the destruction of species with a creative force. For this, Lyell proposes that species are created as single individuals or pairs (as needed), "at such times and in such places as to enable them to multiply and endure for an appointed period, and occupy an appointed space on the globe" (Lyell, 1832, p. 2:129). The production of novel species (presumably divine, although Lyell is tacet on the details here), at a slow and steady pace, is a thus an admissible, if difficult to observe, cause of change, in just the same way as erosion or subsidence are in geology. In areas where geographic isolation occurs, this theory will even will reproduce the appearance of "centres of creation"—areas like South America which appear to have a larger share of tailor-made organisms, because the species being created there at the usual rate are unable to migrate away from the site of their creation (Lyell, 1832, p. 2:131). This cause being added to our uniformitarian arsenal, we have all that we need, Lyell thinks, in order to account for the phenomena we observe—and without proposing anything like Lamarck's radical hypotheses concerning species transformation.

The young Darwin, therefore, starts his work on the species question from a theory of the earth that does its best to downplay the role of chance. It is the fixed and uniform character of natural law that is on display here, the finely tuned balance of the economy of nature propagating itself steadily onward into the future. While new species are assumed to be divinely created, this, too, is subsumed under the broader banner of uniformitarian causation—there is no caprice or arbitrariness in their creation. Darwin's own theorizing for many years leaves this minimization of chance broadly intact, even as he begins to modify some of Lyell's core tenets. His first departure comes as he attempts to integrate his geological and biological findings from the *Beagle* voyage—particularly patterns of extinction which seem to have taken place in the absence of the kinds of environmental changes that Lyell emphasized—with Lyell's account of how species are destroyed. Drawing an analogy between species and individual organisms, Darwin considers a picture on which species are not only continually created like individuals, but also have fixed, maximum durations or lifespans just as individual organisms do (Hodge, 1983). This is, however, only an effort to square Lyell's overall view with the South American fossil record as Darwin interprets it. Darwin has yet to loosen his grip on the perfection and balance of the economy of nature, nor to make any further room for the play of chance:

> If the existence of species is allowed, each according to its kind, we must suppose deaths to follow at different epochs, & then successive births must repeople the globe or the number of its inhabitants has varied exceeding[ly] at different periods. — A supposition in contradiction to the fitness which the Author of Nature has now established.[a]

(Darwin, 1835, fol. 2v)

Darwin is thus still a committed, if at this point slightly heterodox, Lyellian, for whom the distribution of species still shows, most of all, the harmonious and roughly

---

[a] For Darwin's letters and notebooks, I preserve his original spelling, punctuation, and emphasis, except in cases where it is genuinely ambiguous. I will cite the notebooks by the standard notebook lettering and pagination; letters are indexed to the excellent work of the Darwin Correspondence Project.

stable character of the natural order. Adding a second cause for the disappearance of species makes no dramatic change in this regard.

Two years later, opening a notebook containing his current ideas on the species question, Darwin writes what is commonly taken to be the first strong evidence of "transformism" or "transmutationism" in his thought, as hypotheses concerning the evolution of species were most commonly then called (at the beginning of Notebook B, Darwin, 1837; though see Hodge, 1983, p. 80, for skepticism concerning the extent to which this work is really "transformist"). While he begins to consider here the notion that species might change over time, he does not yet alter, in any significant way, his philosophical approach—he is still on the hunt for regular, clockwork "laws of life," as Lyell called them, that would fulfill the same role as Lyell's dual causes for species creation and destruction. Drawing his now-famous, first ever "tree of life" diagram, he realizes that a tree-like structure in which species give rise to similar species by the accumulation of small variations, combined with a principle of divergence over time driving clusters of organisms on the tree apart, will produce the pattern of similarities and differences, species and higher taxa arranged as "groups within groups," that we have come to expect from taxonomic study since Linnaeus.

> *Hence if this is true, [it will be the case] that the* greater the groups the greater the gaps *(or* solutions *of* continuous structure*) between them. – for instance, there would be [a] great gap between birds and mammalia, Still greater Vertebrate and Articulata, still greater between animals & Plants.*
>
> **(Darwin, 1837, pp. B42–3)**

Even this skeletal process of differentiation and divergence already bears some of the features of Darwin's mature works. It proceeds, he says, from three elements which mirror his later presentation of evolution by natural selection—from "infinite variations," from the species "all coming from one stock," and from "obeying one law," here the principle of divergence (Darwin, 1837, p. B43). But the details, of course, are still unspecified (in particular, the mechanism of natural selection as the driver for adaptation is nowhere to be found), and Darwin worries whether or not there will be sufficient evidence for his newly fledged picture. "Heaven know[s] whether this agrees with Nature," he writes. "*Cuidado*" (Darwin, 1837, p. B44).

But even as Darwin begins to more seriously move away from the views of Lyell, embarking upon a project to construct a picture of the natural world in which observed taxonomic structure is a product only of the regular interaction of the laws of nature, he still is not making much, if any, room for chance. The passages above were written around the middle of July 1837, and as he writes the remainder of the B notebook over the ensuing year, all its uses of "chance" refer only to something like a law of large numbers. (To take just one example, at two points Darwin considers how unlikely it is that any currently extant organism will have offspring still living dozens of generations into the future (Darwin, 1837, pp. B41, B146).) He then enters a long period of slow theoretical work, in which it is difficult to locate any major landmarks (Hodge, 2009, pp. 53, 64). His focus widens during this time, as he starts to consider the impact of the evolution of species on the understanding of humans, including our mental and behavioral faculties, and on religious doctrine. He also begins thinking about how to package what he now calls "my theory" as a publishable piece of public science, re-reading Sir John Herschel's *Preliminary Discourse on the Study of Natural Philosophy* in October of 1838 and taking from it a wealth of advice on how to build a causal theory that would satisfy all the tenets of the Newtonian philosophy of science of his day (Pence, 2018). Chance makes one brief appearance here, but only in the "metaphysical" notebook M, as Darwin uncharacteristically grapples with the question of free will. "[W]e may easily fancy there is" such a thing as free will, he writes, "as we fancy there is such a thing as chance. — chance governs the descent of a farthing, free will determines our throwing it up. — equall true the two statements" (Darwin, 1838a, p. M27). Not exactly an attempt at offering a detailed analysis of the concept, nor indicative of it being particularly important to his thought.

A month later, however, we see that chance remains a conundrum. Darwin is contemplating the transmission of habits across generations—important, he thinks, for the understanding of sexual behavior in humans, a topic to which he would return at length in the *Descent of Man*—and he notes that the distinction between habits and instincts parallels nicely that between two sources of variation in heredity:

> *An habitual action must some way affect the brain in a manner which can be transmitted. – this is analogous to a blacksmith having children with strong arms. – The other principle of those children, which* chance? *produced with strong arms, outliving the weaker ones, may be applicable to the formation of instincts, independently of habits.*

(Darwin, 1838b, p. N42)

That is, in the same way that heredity comprises both the inheritance of acquired characters from parents and the generation of novel characters by chance, we should see in species such as humans (with complex patterns of behavior) both the inheritance of acquired behavioral habits as well as the generation by chance of behaviors produced by novel instincts.

One can, however, almost feel the revulsion with which Darwin has underlined the word "chance" in the passage above, which is emblematic of his approach toward the concept throughout this early period. He realizes, it seems, that the generation of variation appears chancy to the outside observer—connected, in some way, to both environmental change and to the habits and characters of parents, to be sure, but by connections too loose and tenuous to be amenable to direct study. But how else to describe it? What kind of explanation could be offered here that was satisfactory, by the lights of both Darwin's biology and his philosophy of science? In some form or another, this question would lurk in the background of Darwin's thought from the mid-1830s until his last works.

It is clear, however, that there is no answer to these questions in Darwin's early notebooks. The role of chance—as with most of Darwin's most important contributions to the study of life on earth—enters gradually, piecemeal, as he puzzles out the correct interpretation of the data he collects from the natural world.

Chance in the *Origin of Species*

Darwin began to seriously formulate his thoughts on the change of species over time in 1842, with a brief *Sketch*, written hurriedly and in pencil, and totaling only 35 pages. He expanded this in a longer *Essay* of 1844, which Darwin left in the care of his wife (along with £400 to cover the costs of its editing and publication), intending to assuage his fears of falling into obscurity due to untimely death from his constant ill health. In both works, the major outlines of Darwin's mature theory as it would later appear in the *Origin* can clearly be seen—some passages from the *Origin* are even found nearly verbatim in the 1842 material. What none of these materials contain, however, is any more clarity on the matter of chance. In the 1842 *Sketch*, Darwin remarks on the "absurdity of habit, or chance ?? or external conditions making a woodpecker adapted to tree" (Darwin, 1909, p. 10), echoing exactly the same disdain for chance that we found in the notebooks of 4 years earlier. Even this brief mention falls away in the *Essay* of 1844, which offers no sustained discussion of the issue.

Darwin then paused his work on natural selection for nearly a decade, writing an encyclopedic work that remains to this day a reference text for the barnacles (prompting one of Darwin's young children to ask a friend where his father "did his barnacles"). He returned to the subject between 1856 and 1858, writing most of a draft for a book—much longer than that which would later become the *Origin of Species*—that he had provisionally entitled *Natural Selection* (Darwin, 1975; Fig. 1.1). Prepared as this draft was with a direct eye toward publication (the first two chapters would eventually go on to become part of *Variation of Plants and Animals under Domestication*, the rest, the *Origin*), Darwin begins finally to tighten up his thoughts on chance.

First, Darwin takes care in this manuscript to distinguish his understanding of the role of chance in natural selection, against the foil of what he calls the "doctrine of chance" or the "fortuitous concourse of atoms"—clearly both references to the kind of random, Empedoclean atomism so artfully mocked by Aristotle (Depew, 2016). Of course, Darwin writes, "no one I should think could extend this doctrine of chance to the whole structure of an animal, in which there is the clearest relation of part to part, & at the same time to other wholly distinct beings" (Darwin, 1975, p. 174). Natural selection is the very opposite of "chance" in this sense which, Darwin thinks, had no hope of explaining the organization of even a single organism, much less the complex interrelationships between organisms within a broader ecosystem. Darwin's worry about drastic misinterpretation here may well have been spurred by the publication in 1844 of the anonymous *Vestiges of the Natural History of Creation* (later revealed to be, as Darwin came to suspect, the work of Robert Chambers; Chambers, 1844), a sloppy and slapdash proto-evolutionary work that was the target of widespread ridicule and derision in polite British scientific circles over the course of the intervening years, including an especially scathing review written by Huxley (Hodge, 1972; Schwartz, 1990). While it is not clear that, in fact, the *Vestiges* actually ascribes any change in the organic world to this random action of chance, some of its interpreters certainly accused it of such. Darwin, on the other hand, always carried

FIG. 1.1

Charles Darwin, in a photograph from around 1854, just before he returned to the *Natural Selection* manuscript draft.

Credit: Photograph by Henry Maull and John Fox, published as the frontispiece to: Darwin, F. (Ed.) 1887. The Life and Letters of Charles Darwin, Including an Autobiographical Chapter. John Murray, London. Public domain image, available on Wikimedia Commons at <https://commons.wikimedia.org/wiki/File:Charles_ Darwin_seated_crop.jpg>.

some sympathy for Chambers's having made evolutionary views at least somewhat more respectable in the public sphere (Darwin, 1876, p. xvii) and did not want his own work to come to the same fate that Chambers's had.

More constructively, he begins to build two roles for chance in natural selection that we will see carried throughout the remainder of his work. First, picking up on a theme that he had only just begun to develop in the earlier drafts, he starts to speak very clearly about natural selection as a *tendency* to improvement. If an organism were born with an improvement over its conspecifics, Darwin writes, "I do not say that it would be invariably selected, but that an individual so characterised would have a better chance of surviving" (Darwin, 1975, p. 214). Hereditary transmission also acts in the same way—so that even if an organism had been favored in the struggle for existence, the best we can say is that it "would in many cases, tend to transmit the new, though slight modification to its offspring" (Darwin, 1975, p. 214).

Second, in an effort to clean up the rather dismayed language we saw throughout the notebooks concerning the role of chance in the initial generation of variations, Darwin tries to clarify his position on this question as well. "[T]he causes, which from their extremely complex nature we are forced generally to call mere chance, which produced the first variation in question [leading to an adaptation] would under the same conditions often continue to act," he writes (Darwin, 1975, pp. 214–215). The generation of variation is governed by causes—no fortuitous conjunction of atoms here—but these are so complex as to be beyond our knowledge, which leaves us with no choice but to refer them to the workings of "chance." At the very least, we can infer that similar conditions will be likely to give rise to similar variations in the future, a kind of vague regularity.

Darwin has thus turned a corner over the course of this drafting process—he has begun to see the way in which chance might be utilized as a positive ally in the construction of natural selection, rather than a mere foil to be resisted. Seeing the generation of variation, its transmission, and the action of natural selection as in some sense chancy processes gives Darwin the room he needs for the messy, haphazard characteristics of the organic world that he so often enlists as evidence for evolution. Natural selection writes no guarantees and offers no perfect solutions—and is nonetheless able to produce adaptations as finely tuned as the eye. What we still do not know on the basis of these short mentions, however, is just what Darwin means when he refers to chance and tendencies. What features in the biological world are grounding these tendencies? For what reasons must natural selection and variation be expressed in this way? In short, while chance has taken on clear roles in Darwin's thought here, it does not yet seem to have taken on a clearer meaning.

For that, we will need to look forward, to Darwin's most famous work, *On the Origin of Species*. The story of how the *Origin* came to be is now fairly well known—in the middle of Darwin's steady progress on his *Natural Selection* manuscript, he received a letter from Alfred Russel Wallace in June of 1858 containing a theory which, to Darwin's eyes, looked nearly identical to his own. Whether or not this is in fact the case—and there are significant differences between Darwin's and Wallace's approaches (Darwin, 1996, pp. 336–337)—Darwin panicked, and the priority of his discovery was established at a meeting of the Linnean Society, in part with an excerpt from the 1844 *Essay*. Darwin then applied himself full-time to the production of his new book, which he now called an "abstract" of the larger volume (which was never completed), and *On the Origin of Species* was published in November 1859.

First and foremost, it is important to note how little changes with respect to the question of chance in the *Origin*, if one is already familiar with the draft materials of the preceding years. Darwin leaves intact both of the primary uses of chance which we saw in the *Natural Selection* manuscript. Regarding variation, he writes that our development of varieties of domestic plants "has consisted in always cultivating the best known variety, sowing its seeds, and when a slightly better variety *has chanced to appear*, selecting it, and so onwards" (Darwin, 1859a, p. 37, emph. added). "Mere chance," he writes later, "as we may call it, might cause one variety to differ in some character from its parents." But, as we saw with his earlier use of the "doctrine of chance," this cannot be the whole story. The sentence continues: "and the offspring of this variety again to differ in the very same character and in a greater degree; but this alone would never account for so habitual and large an amount of difference as that between varieties of the same species and species of the same genus" (Darwin, 1859a, p. 111). Something more is needed—namely the combined action of natural selection and Darwin's principle of divergence, which drives species and groups apart by encouraging them to seize on unclaimed places in the economy of nature (Kohn, 2009; Pence and Swaim, 2018).

The same goes for the description of natural selection as a tendency rather than a law free of exceptions. A variation "in any degree profitable to an individual of any species…will tend to the preservation of that individual, and will generally be inherited by its offspring" (Darwin, 1859a, p. 61). Due to the constant struggle for survival in nature, "individuals having any advantage, however slight, over others, would have the best chance of surviving and of procreating their kind" (Darwin, 1859a, p. 81). We get a succinct combination of both these senses just a few paragraphs later. "[E]very slight modification, which in the course of ages chanced to arise, and which in any way favoured the individuals of any of the species, by better adapting them to their altered conditions, would tend to be preserved" (Darwin, 1859a, p. 82).

This much takes us back to where we were in the *Natural Selection* manuscript. But Darwin goes further here. To open a chapter on the laws of variation, which immediately follows his discussion of natural selection, Darwin writes:

> *I have hitherto sometimes spoken as if the variations – so common and so multiform in organic beings under domestication, and in a lesser degree in those in a state of nature – had been due to chance. This, of course, is a wholly incorrect expression, but it serves to acknowledge plainly our ignorance of the cause of each particular variation.*
>
> **(Darwin, 1859a, p. 131)**

He then discusses the impact of what few known causes of variation he believes he has solid evidence for—with particular focus on the action of changed conditions of life (such as climate, nutrition, and so forth) on the reproductive system.

Here, then, we have Darwin's first efforts at describing what it is that he actually means when he refers to "chance"—at least in the case of variation and, by extension, in the description of natural selection as a tendency.[b] He appeals to one of the oldest interpretations of chance phenomena: chance as subjective ignorance. Chance, when we are forced to describe circumstances in those terms, should be taken only as an indication of our ignorance of the true causes at work in the case at hand. We might or might not be able to eventually obtain the knowledge required to remove any references to chance (Darwin leaves his options open here, though we will see

---

[b] It is uncertain whether Darwin even recognizes as a philosophical problem—as any philosopher of science would today—that he refers to natural selection as a tendency (as opposed to chance variation, which he thought required significant elaboration). As we will see in Chapter 3, this is a problem that would be quickly taken up by biologists in the biometrical school.

that he returns to the question some decades later), but what matters at the moment is that we are only able in the broadest, most general outlines to describe the kinds of causes that lead organisms to vary, and in particular have no possibility of identifying, in the case of a single variation, what causes might have led to its appearance. These variations thus arise "due to chance." Chance, then, is on this view a stand-in for a subjective fact about us as scientific theorists—it is a way in which to describe our inability, in particular places and times, to know the relevant underlying causes.

That this would be Darwin's approach to chance is in some sense unsurprising. While Darwin drew upon influences across the intellectual spectrum, as well as his vast travels, at the end of the day he could not escape his training as a nineteenth-century British man of science. The archetype for scientific knowledge that he was steeped in remained that of Newton, with regular, ordered, lawful Newtonian gravitation still held up as the paradigmatic instance of scientific success. The extensive tradition of British nineteenth-century philosophy of science (unfortunately now too often neglected) only confirms this fact. To draw from just one example that was particularly important to Darwin, Sir John Herschel's *Preliminary Discourse on the Study of Natural Philosophy* (1830) places Newtonian science at its core. Indeed, as has been argued by a variety of commentators (see, e.g., Bolt, 1998, p. 287), one of Herschel's primary goals is to explore the limits of Newton's injunction against the use of hypotheses (*hypotheses non fingo*, or "I do not frame hypotheses," he had famously written in the General Scholium to the *Principia*). The dramatic and repeatable success of the wave theory of light—precisely the kind of thing that, one would have thought, was inadmissible by Newton's strictures—formed, therefore, a significant problem for anyone interested in advancing a Newtonian philosophy of science, and Herschel's work dedicates significant space to elaborating a version of Newtonianism that could meet this challenge. Nonetheless: Newton remains the standard to reach.

Darwin saw himself as a card-carrying member of this tradition. The *Origin* was written, indeed, in precisely the way that one would have expected had Darwin been adhering to Herschel's standards for the structure and argumentation of a scientific text (Pence, 2018). It is thus entirely predictable that Darwin saw his own work as advancing a broadly Newtonian, regularity-driven picture of the natural world. We do not fully know the content of these regularities, especially in the cases of heredity and variation, but this is no reason to believe that they are not in fact present.

Darwin's critics and commentators were not convinced. Most painfully for Darwin himself, Herschel was particularly unimpressed. Darwin wrote to his mentor Lyell, just a few weeks after the *Origin* was published, that "I have heard by round about channel that Herschel says my Book 'is the law of higgeldy-pigglety.'—What this exactly means I do not know, but it is evidently very contemptuous.—If true this is great blow & discouragement" (Darwin, 1859b). The general idea, however—that even the limited and circumscribed uses of chance that Darwin had allowed to remain within the evolutionary process were still far too much for a respectable scientific theory to bear—has been a common refrain in responses to the *Origin*, from Darwin's day down to our own. The review of the *Origin* by Fleeming Jenkin, which caused Darwin to make a number of changes in later editions (Bulmer, 2004; Gould, 1985) accused Darwin and his supporters of "the vague use of an imperfectly understood doctrine of chance" (Jenkin, 1867, p. 289). To hearken forward to the topic of the next three chapters, the American philosopher and psychologist James Mark Baldwin wrote in his report on the 1898 meeting of the British Association for *The Nation* that W. F. R. Weldon's presidential address to the zoological section "rescued so-called 'chance' from its disrepute" (Baldwin, 1898, p. 274). More broadly, the entire philosophical approach that Darwin had employed came in for resounding criticism. In an anonymous review of the *Origin*, the geologist Adam Sedgwick wrote that "I must in the first place observe that Darwin's theory is not *inductive*,—not based on a series of acknowledged facts pointing to a *general conclusion*,—not a proposition evolved out of the facts, logically, and of course including them. To use an old figure, I look on the entire theory as a vast pyramid resting on its apex, and that apex a mathematical point" (Sedgwick, 1860, p. 285).

For his part, Darwin was never quite sure how to deal with the animosity which he so regularly received on this score. He remained, despite all the criticism, convinced that he had followed Herschel's Newtonian-sanctioned methodology. He proposed a novel cause for the origin and diversification of species, and demonstrated both that this cause could be efficacious in the wild and that, if it were, it would lead to a unified explanation of a wide variety of biological, embryological, and paleontological phenomena. The year following the publication of the *Origin*, Darwin writes, exasperated, to his long-time colleague, the botanist J. S. Henslow. Echoing Herschel's insistence on the validity of the wave theory of light, he laments that he would like to ask his critics "whether it was not allowable (& a great step) to invent the undulatory theory of Light—ie hypothetical undulations in a hypothetical substance the ether. And if this be so, why may I not invent hypothesis of natural selection… & try whether this hypothesis of natural selection does not explain (as I think it does) a large number of facts" (Darwin, 1860). Chance, insofar as it features in this methodology for Darwin, is simply a reference to our ignorance about the way that this process works in the fine details—an ignorance which, assuredly, is equally well present in the case of the wave theory of light.

Darwin is thus doing his best here to square a fairly challenging conceptual circle. The rigid, pre-determined universe of Newton and the fluid, accidental, adaptive universe of evolutionary theory are not the most sympathetic of bedfellows. Darwin has, it seems, simply decided that the tension concerning the use of chance that we saw running just beneath the surface since the early notebooks is in fact not a tension at all—or, at least, not any more problematic than the other various conceptual issues that he resolved in the *Origin* and would continue refining over the remainder of his life. Given that, however, it is no great shock to learn that Darwin had, as Galton put it at the beginning of the chapter, "no liking for statistics." For chance is not the sort of concept to be reasoned with, as statistical inquiry would have it. Rather, it is the sort of concept to be reasoned around or reasoned away. The remaining time that Darwin would devote to chance in the rest of his works, therefore, is by and large in this vein. As we will see, Darwin devotes space both in his later discussions of variation and in his less-read book on orchids to the project of making more precise the ways in which apparent chance can be separated from more genuine chance, and (his preferred option) how apparent chance might be dissolved.

Chance after the *Origin*

If one compares the final (and, as we must recall, rushed) edition of the *Origin* with its draft materials, one of the most important differences is the published version's near-complete lack of citations and comparatively scant invocation of examples. Darwin had intended to prepare a work that introduced each of his major claims with support from a vast array of real-world data, which he had been painstakingly gathering from his correspondence with a worldwide network of naturalists, breeders, and horticulturists, among others. The quick publication of the *Origin* meant that Darwin knew he lacked the time for such comprehensive citation in the as-yet unwritten portions, and so he made the painful decision to excise much of this work.

One of his central goals, however, in the years following the publication of the *Origin*, was to prepare and publish these insights, which would cement, he believed, the broader case for evolution by natural selection (Fig. 1.2). To take only a few

FIG. 1.2

Charles Darwin's study at Down House in 1882, as it appeared shortly after his death.

Credit: "Darwin: The Study at Down," engraving by Haig, A.H., 1882. Scan From the Wellcome Library, licensed under CC-BY 4.0, available on Wikimedia Commons at <https://commons.wikimedia.org/wiki/ File:Darwin;_The_Study_at_Down_Wellcome_L0025093.jpg>.

representative examples, he published books describing the methods of fertilization in orchids, the methods of action of insectivorous and climbing plants, the impact of earthworms on their habitats, and an exhaustive catalog of the extent of domestic and natural variation in a vast number of species. Much ink has and still could be spilled describing the relationship between all of these various studies and Darwin's grand project to defend and promote his theory—Darwin's American friend and frequent correspondent Asa Gray wrote of the orchid book that it was a "beautiful *flank*-movement" in the argument for natural selection, a sentiment with which Darwin agreed, calling it his "chief interest" in writing the work (Darwin, 1862a; Gray, 1862). For my purposes here, I will focus on the discussion of chance as it appears in *Orchids* and *Variation*.

The recognition of *Orchids* as a pivotal work for understanding Darwin's thoughts on chance is owed in large part to the work of John Beatty (2006), who in turn credits James Lennox for discovering the central example. Chance is not the work's central topic; Darwin's primary target is to provide further evidence for another claim that he makes in the *Origin*. He spends an entire section of that work's fourth chapter emphasizing the extent and importance of variation for species in the wild, making a case for the general advantage of sexual reproduction in every species with separate sexes. Even in the case of hermaphroditic organisms, he writes, "I am strongly inclined to believe that…two individuals, either occasionally or habitually, concur for the reproduction of their kind. […] We shall presently see its importance [of intercrossing, as providing raw material for natural selection], but I must here treat the subject with extreme brevity" (Darwin, 1859a, p. 96). His *On the Various Contrivances by which British and Foreign Orchids are Fertilized by Insects* is designed to fill in a part of this argument, with an incredibly detailed study of a fairly large and very often hermaphroditic group—hence one that might be expected to cause trouble for his insistence on frequent crossing. Despite widely varying morphologies and a vast array of insect pollinators, Darwin says, "hardly any fact has so much struck me as the endless diversity of structure,—the prodigality of resources,—for gaining the very same end, namely, the fertilisation of one flower by the pollen of another" (Darwin, 1862b, pp. 348–349). With only one exception, he believes, every single orchid at least occasionally, and usually either frequently or always, is fertilized by insects bringing pollen from one plant to another.

In the spirit of Gray's "flank movement" remark, however, Darwin sprinkles in several arguments which connect his observations of orchids to other parts of his broader defense of evolution. For example, he offers a fairly extensive discussion of homology and systematics for the orchids—writing that "if, indeed, he [the 'student of nature'] should care to see how much light, though far from perfect, homology throws on a subject, this will, perhaps, be nearly as good an instance as could be given" (Darwin, 1862b, pp. 288–289). The appearance of "chance variation" in the orchids also gets its turn in the limelight, and it is this that makes it important for our discussion here.

Consider the example of *Malaxis paludosa* (today's *Hammarbya paludosa*, the "bog orchid"), the same discussed by Beatty (2006). One petal of orchids is, usually, that on which insects land to consume nectar and, in the process, transport pollen. For many orchid species, this petal is initially at the top of the orchid when the flower first opens, "but assumes its usual position as the *lower lip* by the twisting of the ovarium," Darwin writes. In this species, however, "the twisting has been carried to that degree that the flower occupies the position which it would have held if the ovarium had not been at all twisted" (Darwin, 1862b, pp. 131–133). Rather than simply remaining stationary, that is, this orchid twists through a complete 360 degrees, to wind up back where it started. This much, then, is remarkable as an indicator that this position of the petals was very clearly acquired as an adaptation to aid fertilization in this species. But Darwin returns to the subject later in a discussion of chance variation. "As all the parts of a flower are co-ordinated, if slight variations in any one part are preserved from being beneficial to the plant, then the other parts will generally have to be modified in some corresponding manner," he writes. "But certain parts may not vary at all, or may not vary in the simplest corresponding manner." Such was the case with our *M. paludosa*. The observed degree of rotation, he writes, "might simply be effected by the continued selection of varieties which had their ovarium a little less twisted; but if the plant only afforded varieties with the ovarium more twisted, the same end could be attained by their selection until the flower had turned completely round on its axis" (Darwin, 1862b, pp. 349–350).

Let's unpack Darwin's reasoning a bit. We take as our initial constraint that natural selection is pushing the population toward what we might call the "upright" final state, from an ancestral plant that, Darwin assumes, was twisting to the "inverted" position, as this is common to the majority of the orchids. Of course, there are two ways to reach "upright" from "inverted"—to untwist, or to keep twisting. Because we don't know the underlying causes of particular variations, we simply can't know what the next step in the process would be. The relevant parts might simply not vary whatsoever, or they could vary in what appears, to us at least, to be the "more complex" direction of variation. Brief glimmers, here, of the way in which chance might be a useful factor in the production of adaptations. But just as importantly, apparently, we are quickly reassured that this is no cause for concern. Chance has not been given inappropriate license, because it remains constrained by natural selection to produce overall better-adapted forms. As he reiterates the matter just a bit further on:

> The more I study nature, the more I become impressed with ever-increasing force with the conclusion, that the contrivances and beautiful adaptations slowly acquired through each part occasionally varying in a slight degree but in many ways, with the preservation or natural selection of those variations which are beneficial to the organism under the complex and ever-varying conditions of life, transcend in an incomparable degree the contrivances and adaptations which the most fertile imagination of the most imaginative man could suggest with unlimited time at his disposal.

(Darwin, 1862b, p. 351)

Again, we see the importance of chance, but equally the importance of keeping chance well circumscribed. "The flowers of Orchids," he writes earlier, "in their strange and endless diversity of shape, may be compared with the great vertebrate class of Fish, or still more appropriately with tropical Homopterous insects, which seem to us in our ignorance as if modelled by the wildest caprice" (Darwin, 1862b, p. 285). But only to those who are unaware of the constraining power of natural selection.

In 1868, Darwin published the first edition of *The Variation of Animals and Plants Under Domestication* (Fig. 1.3). As mentioned above, this book was primarily assembled from the first two chapters of material he had written for the large book manuscript between 1856 and 1858. Unfortunately, these sections of the original manuscript are no longer extant, making it impossible for us to determine how many of the positions he takes in the *Variation* are a product of his mature views (in particular, those he developed as he responded to criticism he received after the publication of the *Origin*) and how many he formed earlier (see Darwin, 1975, pp. 9–12). It offers an elaboration of the material that Darwin had briefly treated in the *Origin*'s first chapter, and is more than twice as long as the entire *Origin* itself—Darwin spent nearly 7 years in writing it, according to his pocket diary (Darwin, 1975, p. 14).

FIG. 1.3

Charles Darwin, photographed by Julia Margaret Cameron in 1869.

Credit: Photograph by Julia Margaret Cameron. Public domain image, available on Wikimedia Commons at <https://commons.wikimedia.org/wiki/File:Charles_Darwin_01.jpg>.

This work constituted Darwin's main effort to explain the notion of chance in the genesis of variation. The task has been made more essential for him, as several critics of the *Origin* have, he writes, "declared that natural selection explains nothing, unless the precise cause of each slight individual difference be made clear" (Darwin, 1875, p. 2:426). Such a requirement would of course be fatal to Darwin's project, and so the several notions of chance at play when he discusses variation receive their fullest treatment here.

We should begin our discussion with a few concepts of "chance" that, while present as a recessive theme in other parts of Darwin's work, appear in a much more robust form in the *Variation*. First, as mentioned briefly above, Darwin makes very explicit that he is entirely aware of something like the law of large numbers. In the context of an argument that it cannot be mere coincidence that causes an individual to inherit characters of its parents, he writes, with the mathematical aid of George Stokes (of fluid dynamics and multivariate calculus fame), that if "in a large population, a particular affectation occurs on an average in one out of a million" individuals, in a population consisting of "sixty millions, composed, we will assume, of ten million families, each containing six members," then "the odds will be no less than 8333 millions to 1 that in the ten million families there will not be even a single family in which one parent and two children will be affected by the peculiarity in question." Absent Darwin's strange double negative here, that is, mere random distribution of a character trait among offspring could in no way account for the "numerous instances [that] could be given, in which several children have been affected by the same rare peculiarity with one of their parents" (Darwin, 1875, p. 1:449). The patterns we see of parents and children carrying the same rare disease are unintelligible on a hypothesis of random distribution of traits in a population, precisely because of the size of the relevant populations and the rarity of the conditions involved; the odds of random distribution producing the phenomena we see are simply too low.

Second, Darwin makes clear here, particularly as the result of letters he was exchanging with Asa Gray, that he intends his picture of variation to be "accidental" in the sense opposed to design. He draws a metaphor, asking us to imagine that an architect goes to assemble a building using only fragments of stone that have been cast from a cliff face, finding there exactly all of the sizes and shapes required. In this case,

> The shape of the fragments of stone at the base of our precipice may be called accidental, but this is not strictly correct; for the shape of each depends on a long sequence of events, all obeying natural laws; on the nature of the rock, on the lines of deposition or cleavage, on the form of the mountain, which depends on its upheaval and subsequent denudation, and lastly on the storm or earthquake which throws down the fragments. But in regard to the use to which the fragments may be put, their shape may strictly be said to be accidental.
>
> **(Darwin, 1875, p. 2:427)**

Variations, then, are exactly the same. They appear "accidental" in the sense that the laws which govern their generation—laws which we can only begin to understand, though which are likely to derive from "the conditions of life to which each being, and more especially its ancestors, have been exposed" (Darwin, 1875, p. 2:241)—are not somehow specially tailored so as to produce variations that would go on to be favored by natural selection. Variation has no such predictive power, nor has it, as Gray hoped to demonstrate, tended to lead organismic form down "useful" evolutionary paths. Again, though, this is not to abandon ourselves to some capricious sort of chance. Rather, it is to emphasize that variations are, in general, "alike in nature and the result of the same general laws," and have gone on to be "the groundwork through natural selection of the most perfectly adapted animals in the world" (Darwin, 1875, p. 2:427–8). No abandonment of the Newtonian project here, even if we are as yet unable to describe how these laws themselves might work.

But what, then, can we say of the causes of variation? We can enumerate a number of cases where the causes can at least be specified at a general level (such as the addition of excess food, or some kinds of grafting), and we can describe a few regularities among cases where variation is generated (such as the likelihood of variations to be correlated among parts, or the likelihood of multiple homologous parts to vary together). But these examples are few and far between, and their basis is generally not well understood, even when the phenomena seem fairly predictable.

The best we can do as a general theory of the causes of variation, then, is to sort them into two types. There are the cases in which environmental conditions "act in a definite manner on the organisation, so that all, or nearly all, the individuals thus exposed become modified in the same manner." This, Darwin thinks, is relatively unusual, and encompasses primarily instances like change in coat length due to direct exposure to a change in temperature. "But a far more frequent result," he writes, "of changed conditions, whether acting directly on the organisation or indirectly through the reproductive system, is indefinite and fluctuating variability" (Darwin, 1875, p. 2:345). That is, for reasons as yet not well known to us, changes in the environmental conditions tend to amplify variability itself, across the organism, such that the increased tendency to vary might express itself in a variety of different ways in different individuals. This indefinite variation is natural selection's primary raw material, and it is the one most clearly grounded in a "chancy" tendency. As Darwin puts it in his summary of the *Variation* in the last edition of the *Origin*, such variations "may be considered as the indefinite effects of the conditions of life on each organism, in nearly the same manner as a chill affects different men in an indefinite manner, according to their state of body or constitution" (Darwin, 1876, pp. 6–7). Nothing strange here, Darwin seems to say—just an instance of the kind of complex dependency of a phenomenon on multiple causes that we are, in fact, used to from our everyday lives. A bit later on, he makes the dependency on complexity even more explicit, writing that "the form of each [being] depends on an infinitude of complex relations, namely on the variations which have arisen, these being due to causes far too intricate to be followed out,—on the nature of the variations which have been preserved…,—and lastly, on inheritance…from innumerable progenitors" (Darwin, 1876, pp. 100–101).

Again, everything here is consistent with the picture Darwin has been constructing all along. He has classified variation to the full extent to which he is able—notably, isolating any and all "non-chancy" sources of variation as far as possible, leaving only the realm of "indefinite variation" as the domain of what we might call genuinely chancy influences. But this, in turn, is placed under the watchful eye of natural selection, and is domesticated by comparison to the action of something as pedestrian as a chill in the night air.

Plausibly, given the paucity of experimental data available to Darwin at this time, combined with his commitment to a solidly deterministic philosophy of science, this is the most detailed and precise way of understanding chance variation that is available to Darwin—and it is the extent of his work on chance. Let's sum up.

Chance, contained

As we have seen, Darwin makes room for a variety of ways in which various concepts of chance might enter into our theorizing about the history of life. Setting aside a few auxiliary notions (accident as opposed to design, or the law of large numbers), the two that wind up being central are natural selection's action as a mere tendency rather than an exceptionless law, and the ignorance of the precise causes of variation that gives rise, in the end, to Darwin's concept of indefinite variation. But every time Darwin introduces such an idea, he immediately hedges and contains it—particularly by deploying natural selection as an ever-present constraint on chance's impact on the actual evolutionary process.

The reason for this back-and-forth motion in Darwin's thought on chance is, I think, fairly clear—his commitment to his deterministic, Lyellian/Herschellian/Newtonian philosophy of science could never allow him to embrace chance more wholeheartedly than this. There is no need, as Johnson (2015) has done, to introduce a hypothesis that Darwin somehow viewed chance as dangerous or subversive, and thus felt the need to bury it within his theory in order to placate critics. This was, rather, a natural demand of his deeper, pre-existing philosophical commitments (commitments he shared with the majority of the nineteenth-century British scientific establishment, not in the slightest bit heterodox), which simply left him with no other path.

And so the stage is set. Darwin has slightly cracked open the door to chance in the evolutionary process, but its full integration into evolution would have to wait for a few more decades, for scholars who lacked Darwin's deep commitment to the Newtonian ideal.

References

Baldwin, J.M., 1898. The British Association. Nation 67, 273–275.

Beatty, J.H., 2006. Chance variation: Darwin on orchids. Philos. Sci. 73, 629–641. https://doi.org/10.1086/518332.

Bolt, M.P., 1998. John Herschel's Natural Philosophy: On the Knowing of Nature and the Nature of Knowing in Early-Nineteenth-Century Britain (Ph.D. thesis). University of Notre Dame.

Bulmer, M., 2004. Did Jenkin's swamping argument invalidate Darwin's theory of natural selection? Br. J. Hist. Sci. 37, 281–297. https://doi.org/10.1017/S0007087404005850.

Chambers, R., 1844. Vestiges of the Natural History of Creation. John Churchill, London.

Darwin, C., 1833. Letter 232—Darwin, C. to Darwin, C. R., 27 Nov. 1833.

Darwin, C., 1996. On Evolution. Hackett, Indianapolis, IN.

Darwin, C., 1835. The position of the bones of Mastodon (?) at Port St Julian is of interest. CUL-DAR42.97–99. Darwin Online, URL. http://darwin-online.org.uk/.

Darwin, C., 1837. Notebook B: [Transmutation of species (1837–1838)]. In: CUL-DAR121. URL, Darwin Online. http://darwin-online.org.uk/.

Darwin, C., 1838a. Notebook M: [Metaphysics on morals and speculations on expression (1838)]. In: CUL-DAR125. URL, Darwin Online. http://darwin-online.org.uk/.

Darwin, C., 1838b. Notebook N: [Metaphysics and expression (1838–1839)]. In: CUL-DAR126. URL, Darwin Online. http://darwin-online.org.uk/.

Darwin, C., 1859a. On the Origin of Species, first ed. John Murray, London.

Darwin, C., 1859b. Letter 2575—Darwin, C. R. to Lyell, Charles, [10 Dec. 1859].

Darwin, C., 1860. Letter 2791—Darwin, C. R. to Henslow. J. S. 8 May. [1860].

Darwin, C., 1862a. Letter 3662—Darwin, C. R. to Gray, A., 23[−4] Jul. (1862).

Darwin, C., 1862b. On the Various Contrivances by Which British and Foreign Orchids Are Fertilized by Insects. John Murray, London.

Darwin, C., 1875. The Variation of Animals and Plants Under Domestication, second ed. John Murray, London.

Darwin, C., 1876. On the Origin of Species, sixth corr. ed. John Murray, London.

Darwin, C., 1909. The Foundations of the *Origin of Species*: Two Essays Written in 1842 and 1844. Cambridge University Press, Cambridge.

Darwin, C., 1975. Charles Darwin's *Natural Selection*; Being the Second Part of His Big Species Book Written From 1856 to 1858. Cambridge University Press, Cambridge.

Depew, D.J., 2016. Contingency, chance, and randomness in ancient, medieval, and modern biology. In: Ramsey, G., Pence, C.H. (Eds.), Chance in Evolution. University of Chicago Press, Chicago, pp. 15–40.

Futuyma, D.J., 2005. Evolution. Sinauer Associates, Sunderland, MA.

Galton, F., 1901a. Letter from FG to KP. 1901-07-04.

Galton, F., 1901b. Letter from FG to KP. 1901-07-08.

Gould, S.J., 1985. Fleeming Jenkin revisited. Nat. Hist. 94, 14–20.

Gray, A., 1862. Letter 3637—Gray, A. to Darwin, C. R., 2–3 Jul. 1862.

Herschel, J.F.W., 1830. A Preliminary Discourse on the Study of Natural Philosophy, first ed. Longman, Rees, Orme, Brown, & Green, London.

Hodge, M.J.S., 1972. The universal gestation of nature: Chambers' *Vestiges* and *Explanations*. J. Hist. Biol. 5, 127–151. https://doi.org/10.1007/BF02113488.

Hodge, M.J.S., 1983. Darwin and the laws of the animate part of the terrestrial system (1835–1837): on the Lyellian origins of his zoonomical explanatory program. Stud. Hist. Biol. 6, 1–106.

Hodge, M.J.S., 2009. The notebook programmes and projects of Darwin's London years. In: Hodge, M.J.S., Radick, G. (Eds.), The Cambridge Companion to Darwin. Cambridge University Press, Cambridge, pp. 44–72.

Jenkin, F., 1867. [Review of] The Origin of Species. North Br. Rev. 46, 277–318.

Johnson, C., 2015. Darwin's Dice: The Idea of Chance in the Thought of Charles Darwin. Oxford University Press, Oxford.

Kohn, D., 2009. Darwin's keystone: the principle of divergence. In: Ruse, M., Richards, R.J. (Eds.), The Cambridge Companion to the "Origin of Species". Cambridge University Press, Cambridge, pp. 87–108.

Lyell, C., 1832. Principles of Geology. vol. 2 John Murray, London.

Ospovat, D., 1981. The Development of Darwin's Theory: Natural History, Natural Theology, and Natural Selection, 1838–1859. Cambridge University Press, Cambridge.

Pearson, K., 1901a. Letter from KP to FG. 1901-04-18.

Pearson, K., 1901b. Letter from KP to FG. 1901-04-30.

Pearson, K., 1901c. Letter from KP to FG. 1901-07-03.

Pearson, K., 1901d. Letter from KP to FG. 1901-07-10.

Pence, C.H., 2018. Sir John F. W. Herschel and Charles Darwin: nineteenth-century science and its methodology. HOPOS 8, 108–140. https://doi.org/10.1086/695719.

Pence, C.H., Swaim, D.G., 2018. The economy of nature: the structure of evolution in Linnaeus, Darwin, and the modern synthesis. Eur. J. Philos. Sci. 8, 435–454. https://doi.org/10.1007/s13194-017-0194-0.

Schwartz, J.S., 1990. Darwin, Wallace, and Huxley, and *Vestiges of the natural history of creation*. J. Hist. Biol. 23, 127–153.

Sedgwick, A., 1860. Objections to Mr. Darwin's theory of the origin of species. The Spectator 33, 285–286.

Sheynin, O.B., 1980. On the history of the statistical method in biology. Arch. Hist. Exact Sci. 22, 323–371.

Sloan, P.R., 1986. Darwin, vital matter, and the transformism of species. J. Hist. Biol. 19, 369–445. https://doi.org/10.1007/BF00138286.

The wonderful form of cosmic order: Francis Galton

I was a good deal drawn by Galton's letter, for it seemed to me that he was still hopelessly at sea with regard to the theory of regression, and if he did not follow the bearing of my January paper on the law of ancestral heredity, who in the world can I expect to?
Pearson, letter to Weldon, 3 August 1898

Our story began with Karl Pearson, in need of a boost in prestige (and finances) for his new journal project in statistical biology, appealing to someone who was widely regarded as one of the leading senior figures (the term "elder statesman" would not be inapt) in the life sciences at the end of the nineteenth century: Francis Galton. Now famous primarily for his role as one of the founding fathers of eugenics (Kevles, 1985), Galton presents a unique interpretive challenge for the 21st century historian or philosopher of science. His early life was more than a bit scattershot. Perusing the table of contents of his autobiography offers us a convenient glimpse of this, as we find an entire chapter on "hunting and shooting," flanked by geographic excursions across Africa and the Middle East (Galton, 1908, p. vii). He only came to an interest in problems of evolution and heredity after reading his cousin Charles Darwin's *Origin of Species* in 1859, at the age of 37. It would be another decade before his first significant "biological" publication, *Hereditary Genius*, appeared (Galton, 1869), and this work, he laments in the preface which he added to the second-edition reprint 23 years later, was not up to the standard of his later efforts (Galton, 1892, p. vii). He dedicated the last years of his life (roughly, from 1900 until his death in 1911) almost entirely to the public proselytization of eugenics (Kevles, 1985, chapter 1). We therefore have only about 30 years' work on problems of heredity and evolution to consider. Galton's views changed significantly over this period, and he was by no stretch of the imagination a particularly good writer or fabricator of philosophical argument—Bernard Norton is quite accurate when he writes (concerning Galton's general notion of "natural ability") that "the argumentation concerning this concept was, as frequently with Galton, very bad, but the concept was powerful if vague" (Norton, 1978, p. 43). Richard Swinburne describes one of his arguments in favor of discontinuous evolution as "quite extraordinarily bad" (Swinburne, 1965, p. 28). Even Pearson himself, in an early lecture on Galton to which we will return in the next chapter, writes that Galton's discussion "is full of 'analogies' which stretch over forms of government, cookery books, and walls of old bricks in a way which seems

The Rise of Chance in Evolutionary Theory. https://doi.org/10.1016/B978-0-323-91291-4.00008-X

to me to contain the maximum of scientific danger, with the minimum of logical advantage" (Pearson, 1889, fol. 8).

At the same time, however, despite Galton's work appearing to be a moving target of dubious quality, his influence was massive. In his biographical memorial for W. F. R. Weldon, Pearson would say that:

> *The next year was to place in Weldon's hands a book – Francis Galton's* Natural Inheritance, *by which one avenue to the solution of such problems [in variation, correlation, and evolution], one quantitative method of attacking organic correlation, was opened out to Weldon; and from this book as source spring two of the friendships [i.e., with Galton and with Pearson] and the whole of the biometric movement, which so changed the course of his life and work.*
>
> **(Pearson, 1906, pp. 13–14)**

Pearson introduces *Natural Inheritance* in his biography of Galton by noting that "it may be said that it created Galton's school; it induced Weldon, Edgeworth, and the present biographer to study correlation," and argued that, despite its errors and primitive statistical technique,

> *no one who studied it on its appearance and had a receptive and sufficiently trained mathematical mind could deny its great suggestiveness, or be other than grateful for all the new ideas and possible problems which it provided. The methods of* Natural Inheritance *may be antiquated now, but in the history of science it will be ever memorable as marking a new epoch…*
>
> **(Pearson, 1930, pp. 57–58)**

Moving into our own day, Ian Hacking famously wrote of Galton that he was a crucial figure in the "taming of chance"—the marshaling of chance as a phenomenon which could be theorized in a positive manner, rather than explained away or minimized. Galton accomplished this, he argued, by virtue of being willing to explain, as opposed to merely describe, phenomena using a statistical law. In Hacking's view, this meant that Galton was the first to see statistical regularities as "in some way autonomous, and not reducible to some set of underlying causes" (Hacking, 1990, p. 181), a critical ingredient in the modern conception of chance. How exactly to spell this out is, of course, a difficult matter (and one to which we will return later), but Hacking has been followed in this claim by a number of other authors (including prior versions of myself; see Ariew et al., 2017, Ariew et al., 2015; Depew and Weber, 1995; Pence, 2015).

This apparent paradox, then, outlines the goal of the present chapter. If we agree with Pearson and Norton that Galton's methods are primitive and his argumentation poor, then how are we to understand the influence he would go on to have? What was Galton doing with the concepts of and roles for chance in the life sciences, both before and in the writing of *Natural Inheritance*, that would lead this book to have such a significant impact on the history of biology?

Of course, in some sense, this question will be impossible to answer—we cannot know precisely what in Galton's picture of the application of chance to the living world would have led to a sea change in the practices of biologists like Pearson and Weldon. But reconstructing Galton's views can offer us our best chance to see through their eyes. When we do, we will find an understanding of the world substantially different from Darwin's—but not, I think, as revolutionary as Hacking would have it. Galton's power is that of suggestion. To borrow Norton's words: "powerful if vague," indeed.

Early sojourns

Galton was immediately gripped by the *Origin of Species*. Later in life, in his autobiography, he would write that its publication "made a marked epoch in my own mental development, as it did in that of human thought generally" (Galton, 1908, p. 287). Galton was, at this time, known primarily as a geographer, having undertaken expeditions across the continent of Africa. His entry to questions in evolution—perhaps unsurprising for a man who would leave his most lasting impression as an ardent eugenicist—was thus through anthropology and ethnography. The first inkling of his interest in evolution can be detected in a pamphlet he read to the London Ethnological Society in 1863 (and had privately printed), entitled *The First Steps Toward the Domestication of Animals* (Galton, 1863). In a clear nod to Darwin's discussions of domestication in the early chapters of the *Origin*, Galton concludes the pamphlet by writing that, rather than ascribing the domestication of animals to intentional effort, either on the part of entire "barbarian" groups or single individuals (after all, such "intellectual" effort would run counter to his other, racist, assumptions), we should think instead that "a vast number of half unconscious attempts have been made throughout the course of ages, and that ultimately, by slow degrees, after many relapses, and continued *selection*, our several domestic breeds became firmly established" (Galton, 1863, p. 17, original emphasis).

This application of evolutionary theory seems to have engaged Galton in a way that few previous topics had. Around a year and a half after he presented his pamphlet, he would publish a two-part article on "Hereditary Talent and Character" (Galton, 1865a, b), picking up right where he had left off. "The power of man over animal life," he writes, "in producing whatever varieties of form he pleases, is enormously great." In this essay, then, "it is my desire to show, more pointedly than – so far as I am aware – has been attempted before, that mental qualities are equally under control" (Galton, 1865a, p. 157). The results here are particularly fragmentary (and make no reference to statistics or chance), but he already looks ahead to "a future volume on this subject," which would become *Hereditary Genius*.

This book, which he published in 1869 and reprinted in 1892, includes some of Galton's first discussions of the "very curious theoretical law of 'deviation from an average,'" which Galton has introduced from the work of "M. Quetelet, the Astronomer-Royal of Belgium" (Galton, 1869, p. 26; Fig. 2.1). It is worth backing up a bit to consider the way in which Quetelet introduces this law—which was at the time known more commonly as the "law of error" and now goes by the (rather normatively loaded) name of the "normal distribution."

FIG. 2.1

Adolphe Quetelet, in an engraving from 1875.

Credit: Public domain, available from the Library of Congress Prints and Photographs Online Catalog,
<https://www.loc.gov/pictures/item/2003665609/>.

Adolphe Quetelet, now justly recognized as one of the pioneers in the application of statistics to social life, nonetheless took what would appear to modern readers as a rather peculiar approach to the problem—one which, as Theodore Porter notes, "won no important converts, and…has never been seen as a viable approach to the human sciences" (Porter, 1985, p. 51). Quetelet's target is a detailed description of (or, perhaps better, the construction of) *l'homme moyen*, the average man—the type from which all other members of society may be seen to depart from as deviations. In his work *On Man and the Development of His Faculties* (of which Darwin owned a copy, though likely never read), he writes:

> *The man that I consider here is, in society, the analogue of the center of gravity in bodies; he is the mean around which oscillate the members of society; he is, if you will, a fictitious being, for whom all things occur in conformity with the mean results obtained for the society. If one looks to establish, in some sense, the basis of a* social physics*, it is him one must consider, without dwelling upon particular cases or anomalies, and without seeking whether some individual may have obtained greater or lesser development in one of his faculties.[a]*

> **(Quetelet, 1835, p. 1:21, original emphasis)**

---

[a] All translations from the French are my own.

The study of the average man has only recently been made possible, he notes, by virtue of our having obtained large amounts of data concerning our societies. Over the course of the book, Quetelet enumerates broad, statistical trends which describe the influence of perturbing factors on the course of human life—the things which might cause each one of us to fail to realize the characteristics of the average man. These range from age and sex differences to differences in place, season, hour of the day, profession, food, affiliation with various political and religious institutions, and so forth. Given the lack of available data on many characters of interest, he adds, "I only present this work as the sketch of a vast painting, of which the frame may only be filled by infinite pains and immense research" (Quetelet, 1835, p. 1:26). But statistics guarantees that such research is possible in principle. "The calculus of probabilities," he writes, "shows that, all else equal, one approaches all the more closely the truth or the laws that one wishes to grasp, the larger the number of individuals encompassed by the observations" (Quetelet, 1835, p. 1:13–14).

What do we find when we collate this data? First, and most strikingly, we find that there is an unexpected constancy to many properties of human populations. "What sad condition of the human race! We can enumerate in advance," Quetelet laments, "how many individuals will dirty their hands with the blood of their fellow men, how many will be counterfeiters, how many poisoners, more or less like we can enumerate in advance the births and deaths that must take place" (Quetelet, 1835, p. 1:10). This, then, raises the issue of why such regularities would hold over time. During the next few decades, Quetelet develops an answer to this question. The laws to which human development is subject do not take any arbitrary form, but rather the very particular shape of the astronomer's law of error, or the normal curve. Indeed, he writes later, "this same law, so simple and so elegant, applies not only to height [his favorite example], but also to weight, to strength, and in general to all physical laws, we even add, to the moral and intellectual laws of man" (Quetelet, 1871, pp. 253–254). As Porter puts it, after some initial reluctance, "the stability of the numbers of moral statistics soon became a great selling point for this new science" (Porter, 1985, p. 61). Why, then, does this law of error apply to so many properties of human populations? Because real human populations are, in essence, constructed from a series of deviations from the ideal man. "Everything thus occurs as if there existed a typical man, from which all the other men diverge more or less" (Quetelet, 1846, p. 142). But not just in man—"one finds that, in organized beings, all elements are subject to vary around a mean state, and that the variations which arise under the influence of accidental causes are governed with such harmony and precision that one can classify them numerically in advance… All is foreseen, all is governed: only our ignorance leads us to believe that everything is abandoned to the caprice of chance" (Quetelet, 1848, p. 17). Statistics becomes necessary as a result of the difficulty of measuring these accidental causes leading to perturbation away from the mean, which are "capricious enough and diverse enough, as we may suppose, that they render impossible, in certain circumstances, the analysis of all the laws of nature" (Quetelet, 1871, p. 274).

Quetelet, then, builds a picture of the social world on which each human population produces a type, and is then built out of the deviation of particular individuals from that type. What did this insight mean to Galton? Eventually, it would mean quite a bit. Nearly twenty years later, Galton would wax poetic (and racist) about the impact of Quetelet's law of error:

> *I know of scarcely anything so apt to impress the imagination as the wonderful form of cosmic order expressed by the "law of error." A savage, if he could understand it, would worship it as a god. It reigns with serenity in complete self-effacement amidst the wildest confusion. The huger the mob and the greater the anarchy the more perfect its sway.*

(Galton, 1886, pp. 494–495)

The earlier Galton of *Hereditary Genius* was, however, much less impressed. Since the deviation of values from the average would be the same in all cases in which the underlying distribution of perturbing causes was the same, he writes that the normal distribution "may, therefore, be used as a most trustworthy criterion, whether or no the events of which an average have been taken, are due to the same or to dissimilar classes of conditions" (Galton, 1869, p. 29). The normal curve, that is, might be a useful way in which to compare different population-level cases to one another, in order to determine whether they had been subjected to the same kinds of conditions. We do not yet find in Galton's work any particularly sophisticated use of statistics or chance.

What we do find, however, is a budding interest in a theory of heredity which might be able to give rise to these observed normal distributions of characters. Darwin had published his initial theory of pangenesis in 1868, just 1 year before *Hereditary Genius* was printed, and Galton spends a full chapter considering its import. He offers nothing but praise. Pangenesis, he writes, "gives a key that unlocks every one of the hitherto unopened barriers to our comprehension of [heredity's] nature; it binds within the compass of a singularly simple law, the multifarious forms of reproduction, witnessed in the wide range of organic life, and it brings all these forms of reproduction under the same conditions as govern the growth of each individual" (Galton, 1869, p. 364).

Perhaps most importantly of all, it is here that Galton for the first time offers what will become one of his most lasting contributions to the future research program of the biometrical school (and which would survive the impending death of the theory of pangenesis which led to its proposal): his account of the importance of ancestry via the connection between currently exhibited characters and those of distant relatives. How ought we to explain the sudden appearance of "sports" that are significantly different from their parents, or "reversions" that call back into existence ancestral characters that had been absent for several generations? On Darwin's view, as Galton describes it,

> *the gemmules of innumerable qualities, derived from ancestral sources, circulate in the blood and propagate themselves, generation after generation, still in the state of gemmules.... Hence there is a vastly larger number of capabilities in every living being, than ever find expression, and for every* patent *element there are countless* latent *ones.*

(Galton, 1869, p. 367, original emphasis)

This distinction between patent and latent characters will persist in Galton's work, and we will see it recur as one of the central phenomena which the biometrical school tasked themselves with explaining.

In addition to its ability to offer us firm grounds for the distinction between patent and latent characters, Galton's love for pangenesis arose from another source: the ease with which it might, at least speculatively, be susceptible to mathematical analysis. "The doctrine of Pangenesis," he writes, "gives excellent materials for mathematical formulæ, the constants of which might be supplied through averages of facts, like those contained in my tables, if they were prepared for the purpose" (Galton, 1869, p. 370). Put differently, he argues, there is no reason that we might not produce "a compact formula, based on the theory of Pangenesis, to express the composition of organic beings in terms of their inherited and individual peculiarities, and to give us, after certain constants had been determined, the means of foretelling the average distribution of characteristics among a large multitude of offspring whose parentage was known" (Galton, 1869, pp. 372–373).

We thus have the nascent basis of a new research program for Galton. In the service of predicting the future average distribution of characters (with such predictions, in turn, often in the service of the aims of eugenics), we should aim for a mathematical formulation of the proportion and transmission of gemmules over time. This, combined with an understanding of the mechanics of latency and patency, would result in a predictive science of inheritance, grounded in a causal description of the way in which gemmules lead to the development of particular characters. Galton was on track, he thought, to essentially resolving the nature of inheritance (Schwartz, 2008, p. 11).

A fatal problem with this picture soon emerged, however—and it was the result of Galton's own research. As he put it, "it occurred to me, when considering these theories, that the truth of Pangenesis admitted of a direct and certain test" (Galton, 1871, p. 395). If the gemmules, before becoming concentrated in the sexual organs, circulate freely in the blood (a reasonable expectation following Darwin's insistence that pangenesis also explains the inheritance of characters acquired over the lifetime of an individual organism), then a blood transfusion should serve equally well as a transfusion of heritable characters. An extensive set of blood transfusion experiments between rabbits, backed by breeding to determine the transfusions' effect on future generations, could thus handily determine whether or not pangenesis worked as advertised.

It did not. No characters were successfully transmitted via blood-borne gemmules. Darwin would protest that while he had argued that there might at least occasionally be gemmules in the blood, it was by no means a necessary feature of the theory that they be always circulating in the bloodstream, but the damage was done. When Galton's experiments were combined with the increasing prominence of August Weismann's introduction of a theoretical and material separation between germ cells as transmitters of heredity and somatic cells as wholly separate from the processes of inheritance, pangenesis was broadly abandoned. For his part, Galton would very quickly begin to deny the existence of any transmission of acquired characters from parent to offspring, although (perhaps out of an excess of deference) he would continue to call his conception of heredity "the theory of Pangenesis with considerable modification, as a supplementary and subordinate part of a complete theory of heredity" (Galton, 1876, p. 330).

Without the grounding of pangenesis—or, perhaps more broadly, the idea that if pangenesis were correct in broad outline, its details would be filled in in good time as a result of further biological research—Galton was forced to develop his own approach to what lay beneath the apparent phenomena of patent and latent characters and their attendant normal distributions. Thus do we move from Galton's early forays to his first genuine theoretical work.

Galton's early theory of heredity

On Friday, February 9, 1877, Galton (Fig. 2.2) delivered a lecture at the Royal Institution, which would later be published as a series of three articles in *Nature* entitled "Typical Laws of Heredity." This work provides the first published evidence of Galton's novel project. He now has begun to realize that statistical descriptions of the phenomena of heredity are not only useful as simple comparisons across populations, but might further contribute in their own right to the understanding of hereditary transmission itself.

He begins with an expanded version of Quetelet's focus on the stability of statistical properties. Not only is the average individual within a population roughly constant over time, as Quetelet had postulated, but further, thanks to the normal distribution, "there will be tall and short individuals, heavy and light, strong and weak, dark and pale, yet the proportions of the innumerable grades in which these several characteristics occur tends to be constant" (Galton, 1877a, p. 492). That is, the structure of deviations around the average member of the population also remains roughly static over time—a position that Quetelet claimed not to endorse, as he believed that human society could be improved by education and social intervention.[b] (For those wondering how Galton's embrace of population-level stasis could account for evolutionary change, we will return to that problem in due time.) Notably, and as we will see when Galton returns to this concern, this is not simply due to like individuals leaving like as offspring—for the offspring produced by any particular organism are just as much governed by the normal distribution as the population at large, leading average parents to occasionally have extreme offspring, and vice versa. Thus, the question is rather: "How is it that although each individual does *not* as a rule leave his like behind him, yet successive generations resemble each other with great exactitude in all their general features?" (Galton, 1877a, p. 492, original emphasis).

---

[b] Of course, given Quetelet's equally strong commitment to the static nature of statistical distributions and their recurrent features year after year, it is not clear whether Quetelet was consistent on this point.

FIG. 2.2

Francis Galton, painted in 1882 by noted portraitist Gustav Graef.

Credit: Painting from 1882 by Gustav Graef, National Portrait Gallery, London. Public domain image, available on Wikimedia Commons at: <https://commons.wikimedia.org/wiki/File:Sir_Francis_Galton_by_Gustav_Graef.jpg>.

To start, Galton needs to make the problem of individual variation suitable for mathematical analysis, recalling that 8 years earlier he had said only that such formulas might be a possible matter for future research. Unfortunately, we shall see here a pattern in Galton's argumentation that will recur throughout his work, and which makes our task in analyzing him unusually difficult. First, Galton begins by making a mathematical simplification:

> *The outline of my problem of this evening is, that since the characteristics of all plants and animals tend to conform to the law of deviation, let us suppose a typical case, in which the conformity shall be exact, and which shall admit of discussion as a mathematical problem, and find what the laws of heredity must then be to enable* successive *generations to maintain statistical identity.*

(Galton, 1877a, p. 493, original emphasis)

Galton proposes that we start by abstracting to a case where all characteristics of all organisms already exactly exhibit a normal distribution, and then ask ourselves how future inheritance might preserve those distributions. This is, perhaps, a legitimate abstraction to make, but it depends on several further pieces of argument, many of which Galton will in fact rarely offer us. First, it raises the issue of why the case in which populations conform exactly to the normal distribution would be the "typical" one. By extension, we are owed an account of how the typical case should be connected to less typical, real-world examples. And finally, we still lack an account of why these processes would be instantiated by the biological world at all. As we will see below, Galton himself recognizes all of these as serious concerns—but solves precious few of them.

As the next move in his lecture, he considers the depiction of the law of deviation itself. Again, in what will become a common Galtonian pattern, he turns immediately to a mechanical model, his quincunx. The device is now fairly well known, being a staple of science museums across the world, but it still offers an interesting window into Galton's thinking. A board is set up with an opening on top, a pattern of pins drilled into it like many copies of the "5" on a die (from which the device's name), and a funnel leading to the top-center of the device (see the left-hand schematic, marked "Fig. 7," in Fig. 2.3). A handful of pellets is added to this funnel, and they cascade down, bouncing off of the pins along the way. With a sufficiently large number of pellets added, the pattern formed at the bottom of the apparatus will invariably be that of the normal distribution.

Galton's explanation of this fact, in turn, mirrors Quetelet's understanding of the generation of the normal distribution in human populations: each of the pellets begins at the same "average" value, is carried down by the same force of gravity, and is then perturbed from that mean by each of the impacts with the pins. In Galton's words,

> [I]n addition to [the constant effects of the mean value and gravity] there were a host of petty disturbing influences, represented by the spikes among which the pellets tumbled in all sorts of ways. The theory of combination shows that the commonest case is that where a pellet falls equally often to the right of a spike as to the left of it…. It also shows that the cases are very rare of runs of luck carrying the pellet much oftener to one side than the other.
>
> **(Galton, 1877a, p. 495)**

We thus have a perfect explanation for the behavior of these pellets. The "average pellet" will rebound off of the pins an equal number of times to the left and the right, and land in the center, directly beneath the funnel—hence the largest number of pellets accumulates there. But "runs of luck," sometimes smaller and sometimes more significant, will occasionally carry some pellets farther to one side or the other, producing on the average the bell-shaped curve we expect from such a device.

With slight modifications, the model can also be used as a demonstration of important points about the relationship between these normal curves as they develop over time. As mentioned above, the persistence of their features is not a result of like organisms leaving like offspring (i.e., of pellets falling "straight down" from one curve to the next). To show this, Galton asks us to consider the behavior of the subdivided quincunx (labeled "Fig. 8" in Fig. 2.3). Imagine blocking the machine halfway up, at point A, and allowing the pellets to collect in the columns, each of which has an independent trap door that we can remove to allow the pellets there to fall onward toward point B. The pattern at A will form a normal distribution, as it is simply a standard quincunx. Consider, then, what happens if we open just one of the small columns at point A. The pellets that drop from this mid-level bucket will form a small normal distribution, centered around the place from which they fell. We know that, in the end, the overall distribution that collects at B will be normal (as the total fall, whether temporarily blocked or no, should still produce the expected total result). But we can also see that while some of the far-right-hand pellets will have started out in far-right-hand bins at A and fallen straight down (that is, some will be extreme descendants of extreme parents), in fact, more of the pellets that end their journey on the far-right-hand side will have started out closer to the center and worked their way back to the outside in the interval between A and B (that is, more will be extreme descendants of average parents)—and this is so simply because there are far more

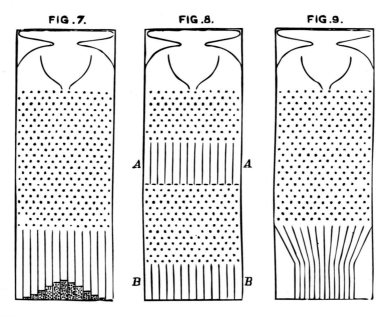

FIG. 2.3

Three versions of Galton's quincunx device, in a figure taken from *Natural Inheritance* (1889).

Credit: Galton, F., 1889. Natural Inheritance, Macmillan, London (Chapter 5), p. 63, figures 7–9. Public domain book, available from the Internet Archive at <https://archive.org/details/galton-francis.-natural-inheritance-1889_202008/page/n3/mode/2up>.

pellets in the central columns to start with than there are pellets near the outer edges. The far-right-hand pellets at the bottom of the device are therefore a mix, with some having arisen from far-right-hand pellets in the "previous generation" and some having moved in that direction from more centrally located positions as they fell.

We remain, however, far from finished. For while the quincunx serves as an effective analog for a derivation in Quetelet's style of the law of error, it does not suffice to explain the appearance of normal distributions in living organisms. In short, the causation at work in the two cases clearly differs. Returning to Galton's lecture,

> [A]lthough characteristics of plants and animals conform to the law, the reason of their doing so is as yet totally unexplained. The essence of the law is that differences should be wholly due to the collective actions of a host of independent petty *influences in various combinations.... Now the processes of heredity...are not petty influences, but very important ones.*
>
> **(Galton, 1877b, p. 512, original emphasis)**

Galton realizes that he cannot in this case help himself to a direct analogy between the behavior of the quincunx and the natural world, precisely because the distribution of small impacts in the case of the pellets (and, more controversially, of the many perturbing influences upon individual people to which Quetelet had pointed) simply have no analog in the case of selection and heredity, as these processes very often have dramatic and wide-ranging effects (up to and including death).

How, then, are we to proceed? We see here yet another microcosm of Galton's questionable argumentative strategy. He performs experiments on peas, in search of at least a "suggestion" for how the processes of heredity could "work harmoniously with the law of deviation, and be themselves in some sense conformable to it" (Galton, 1877b, p. 512)—that is, how the various tendencies of selection or reversion might work together in such a way as to produce the intra-generational statistical stability that Galton was seeking to explain. He then introduces two forces, which he calls family variability and reversion, which he argues can combine to produce the observed results in his peas. Rather than describe the results in biological terms, however, he once again turns to a mechanical model. Family variability, or variation at each generation, is the process analogous with the simple quincunx, in which pellets enter at the top and spread as they move to the bottom. This, Galton realizes, cannot be the sole process at work, for in this case "the dispersion of the race would indefinitely increase with the number of the generations" (Galton, 1877b, p. 513). Each pass through a quincunx like the left-hand side of Fig. 2.3 increases the variance of the population, contrary to the behavior of constancy that Galton hopes to describe.

We thus need to introduce another process, reversion, to counteract this increase in variance. To demonstrate it, consider the right-hand quincunx, labeled "Fig. 9" in Fig. 2.3. Here we have, quite simply, an arbitrary and direct way to counteract the increase in variance caused by family variability, by squeezing the normal curve in the horizontal direction as the pellets fall toward the bottom of the machine. The real process of inheritance in the peas must instantiate something like these two sub-processes in turn, Galton reasons, in order to produce the observed phenomena. Natural selection gets only a brief mention near the end of the lecture, and then only when the selective optimum happens to be the same mean at which the population already finds itself, selection tending, he thinks, to favor the state "in which the demerits of excess or deficiency are most frequently balanced" (Galton, 1877b, p. 514).

We thus know, Galton argues, "the way in which the resemblance of a population is maintained." As noted above, Galton realizes that these are special cases—indeed, we can easily imagine instances in which reversion or selection were centered around a different mean. But these are nonetheless the "typical laws," and while "they may never be exactly correct in any one case…at the same time they will always be approximately true and always serviceable for explanation" (Galton, 1877c, p. 532).

It is important to stop here and take stock, not least because this last sentence has formed a pivotal piece of evidence for authors such as Hacking, who place Galton at the center of the "taming of chance." Galton's decision, Hacking argues, to assert that a purely statistical law could be "serviceable for explanation" marks a dramatic departure from the prior descriptive use of statistics, a departure which signals a sea change in both the use of chance and in the philosophy of science more generally.

I am skeptical. Galton seems clearly to be of two minds about the role that these statistical descriptions are to play. On the one hand, he writes that the statistical laws will be "serviceable for explanation." But simultaneously, as a result of the lack of connection with the biological details, he emphasizes the extent to which they are a special case, acknowledges a variety of ways in which they could fail to hold, and even in the cases where populations might follow them, writes that the "reason of their doing so [i.e., being in conformity with the law of deviation] is as yet entirely unexplained." Thus, while Galton does certainly make more use of the statistical laws than anyone who has come before him, and uses them to infer "suggestions" about the kinds of processes which might underlie them, at this stage he has only mechanical models to explicate their method of action. In the absence of something like the theory of pangenesis—one can almost feel Galton's disappointment that the biological details didn't allow pangenesis to directly account for the processes of variability and reversion—Galton is firmly aware that he has more work to do, and in fact sets himself about the task in the ensuing dozen years during which he continues to explore these problems. Were the difficulty merely solved by the demonstration of a few normal curves and some mathematical models which generate them, Galton—and those biologists like Pearson and Weldon who came after him within a self-identified "Galtonian" school—would have felt no need for the filling in of further details.

Galton's commitment to and interpretation of chance and statistics is, therefore, more nuanced than Hacking's picture (and those that derive from it mentioned earlier, like Ariew, Rice, and Rohwer, as well as my own prior work) would permit. In order to unpack it, then, we need to delve more deeply into his later work, as he molds the basic view here into the mature picture that would appear in *Natural Inheritance*.

Toward a novel theory

Of course, here we are met with another problem. Galton clearly lacks the experimental data required in order to flesh out the process of inheritance in what we would today call biochemical terms. What, then, does Galton believe he can do in order to close the gap between his hypothesized (and mechanically modeled) statistical processes and the real processes at work in the biological world? Galton utilizes two primary resources. First is a connection between the observed statistical patterns and a newly proposed mechanism of inheritance, which owes much to, but attempts to avoid the problems with, Darwin's pangenesis. The second, in turn, is a set of analogies for how the particles of parents might relate to particles of offspring—allowing Galton, as best he is able, to unite his new statistical work on regression with hypotheses about the biological processes that give rise to it.

We return to the story in 1885 and Galton's presidential address delivered before the Royal Anthropological Society. Galton has in the intervening years struggled to simplify his understanding of the statistical underpinnings of his 1877 data (another reason, I think, that we must be careful not to ascribe too much to the state of affairs in that earlier lecture), and in fact we see no invocations of quincunx devices here. "I was then blind," he writes, "to what I now perceive to be the simple explanation of the phenomenon, so I thought it better to say no more until I should obtain independent evidence" (Galton, 1885, p. 1207). The evidence which he collects is that of measurements of height in people, a set of data that he would continue to exploit until the end of his career. For the first time, he phrases his conclusions in terms of what we now call a coefficient of regression:

> An analysis of the records fully confirms and goes far beyond the conclusions I obtained from the seeds. It gives the numerical value of the regression towards mediocrity as from 1 to ⅔ with unexpected coherence and precision, and it supplies me with the class of facts I wanted to investigate—the degrees of family likeness in different degrees of kinship, and the steps through which special family peculiarities become merged into the typical characteristics of the race at large.
>
> **(Galton, 1885, p. 1207)**

While the full details of Galton's work on the method of regression would take us much too far afield for this volume (see Stigler, 1986, pp. 283–296 for a technical reconstruction of Galton's use, and frequent misuse, of regression), it's worth elaborating why Galton believed this single value was the most important piece of data to arise from his new analysis.

Very roughly, Galton's value of ⅔ for the regression coefficient implies that if the deviation of one particular organism from the mean value for the population is x, we should estimate that the deviation of that organism's offspring from the population mean value will be, on the average, ⅔x. Put differently, offspring will be around two-thirds as "extreme" as their parents. This is clearly similar to the process of reversion that Galton was gesturing at in 1877 with his simplistic "squeezing" quincunx, but it is now significantly more mathematically precise—even with the primitive nature of

Galton's methods (for instance, he usually estimated these coefficients just by look-ing at graphs). Stature, as Galton notes, forms a particularly good source of statistical data. It is readily measured (though some error will be introduced by the presence or absence of shoes), more or less invariable during adult life, often recorded or remem-bered even for now-deceased extended family members, and very little otherwise selected for in human life (Galton finds no evidence, for instance, of any preferential mating with regard to height in his data).

Further, Galton has started to build an underlying explanation for this statistical pattern, and it finally allows us to connect back to his earlier invocation of latent and patent characters. He writes that not only is the offspring regression coefficient demonstrated by the observed facts, but "we ought to expect filial regression, and…it should amount to some constant fractional part of the value of the [average parental] deviation" (Galton, 1885, p. 1209). Why? Because of the fact that individuals inherit latent and patent characters both from their parents and their more distant ancestors:

> The child inherits partly from his parents, partly from his ancestry. Speaking gen-erally, the further his genealogy goes back, the more numerous and varied will his ancestry become, until they cease to differ from any equally numerous sample taken at haphazard from the race at large. Their mean stature will then be the same as that of the race; in other words, it will be mediocre. Or, to put the same fact into another form, the most probable value of the mid-ancestral deviates in any remote generation is zero.
>
> **(Galton, 1885, p. 1209)**

Galton's argument is, effectively, as follows. At any extremely remote time in the past, the ancestors of any particular organism will not differ from the average for the population at large in any way whatsoever, as they will be an effectively random sample taken from that population. Their deviation from the "racial type" would thus be, by definition, zero. On the other hand, if an organism simply were identical to its parents, the ratio of its deviation to that of its parents would be one—every organ-ism would be exactly as extreme (or not) as its parents were. But since inheritance is the product of the mixture of these two influences—some influences from parents, and some influences from more remote ancestry—we should find that the regression coefficient that Galton has measured lies somewhere between the two extremes, that is, somewhere between zero and one. It is, Galton writes, "a perfectly reasonable law which might have been deductively foreseen" (Galton, 1885, p. 1210). (Unable to resist his tendency to produce physical models, however, Galton adds that "I have made an arrangement with one movable pulley and two fixed ones by which the probable average height of the children of known parents can be mechanically reck-oned…" (Galton, 1885, p. 1210).)

Further, it is in this lecture that Galton begins to offer a replacement for the understanding of the process of inheritance itself that he lost when he abandoned pangenesis. Insofar as different traits seem to be inherited (and exhibit patterns of regularity) independently of one another, Galton claims that it is evident we should thus conceive of inheritance as "particulate," in the sense that organisms develop by inheriting a bundle of particles from their parents, with the particles governing the production of one feature descending from particles which governed the production of that feature in the parents. What we require, then, is a description of the way in which the particles responsible for the development of an organism are related to the particles that would be inherited by that organism's offspring. To that end, Galton deploys an analogy that he first used in *Hereditary Genius*, and that had by now become his favorite metaphor for this relationship: the constitution of parliamentary bodies. The connection between the characters of a parent and those of an offspring should, he writes, be roughly like that between the representative government of a nation and that of one of its colonies. Certain rough affinities would be seen, no doubt because the voting members of the colony would be roughly representative of the voting members of the homeland—but very slight alterations in that sampling process could engender very large overall differences to the colonial government, just as small changes to a parliament with nearly balanced parties can also result in significant shifts in governmental policy. To see this analogy in further detail, however, we will have to look forward to the pivotal work with which we began our narrative: Galton's *Natural Inheritance*.

Before moving on to his mature view as presented there, I should stop to make clear that Galton has in no way ceased to care about the underlying biological facts in the case of stature—again, I think, evidence that the statistical claims at issue here are not properly "autonomous." Rather, stature offers us a peculiar case in which we are in fact able to offer a Quetelet-style explanation of why it is that stature adheres to the normal curve so precisely. This is, Galton writes, "due to the number of variable elements of which the stature is the sum" (Galton, 1885, p. 1208). Unlike in many other cases of heredity, where Galton had worried about whether or not each contributing cause could really be understood as a "petty influence," he argues that when considering stature, we can indeed help ourselves to such an underlying causal structure of petty influences (of the lengths of bones, strength of tendons, and so on), and hence have saved ourselves from needing to offer an explanation of why the data would appear to follow a normal curve in the first place.[c]

Galton's *opus magnum*

The lecture of 1885 thus has all of the ingredients that Galton would require for the presentation of his mature theory of inheritance in what is widely recognized to be his crowning achievement: *Natural Inheritance* of 1889. He combines here the

[c] Note that, unlike Ariew et al., I do not find evidence that Galton argued, in general, that all heritable characters were thus distributed, and hence met the "minimal material requirements" of a normally distributed variable, i.e., being composed of numerous and independent causes (Ariew et al., 2017, p. 64). Rather, Galton offers us an argument for why height — as opposed to his earlier data on peas from 1877, where no such argument was available — would in fact constitute such a system, and hence would not require further explanation of its normality. This is a special case, not a general feature.

fundamental particulate nature of heredity, a defense of both the utilization of chance and statistics to assess inherited characters and the relationship between statistics and the underlying biological causes, as well as a discussion of the influence of natural selection. Let's consider each in turn.

Particulate inheritance

Unlike his earlier works, where the question of the nature of inheritance takes a clear back seat to the statistical concerns, Galton—looking, I suspect, to put his work in a more coherent order by beginning with the underlying process of inheritance— introduces the book with an examination of particulate inheritance as the fundamental mechanism of heredity. Referring to the pattern of inheritance of traits as discrete units from our parents and other ancestors that he discussed in the prior lecture, Galton writes (anticipating another political metaphor to which he will return) that

> It would seem that while the embryo is developing itself, the particles more or less qualified for each new post wait as it were in competition, to obtain it. Also that the particle that succeeds, must owe its success partly to accident of position and partly to being better qualified than any equally well placed competitor to gain a lodgement.

(Galton, 1889, p. 9)

The circumstances under which this competition occurs, however, are "small and mostly unknown" (Galton, 1889, p. 9).[d] The vocabulary with which we should describe inheritance, then, is of this collection of particles, containing the ability to build more traits than are in fact manifest in the adult organism. Given that (one is tempted to say ontological) commitment, we need now endeavor to recast what little knowledge Galton had obtained about the basic processes in particulate inheritance—especially about the two processes of family variation and reversion, and about the distinction between patent and latent characters—in the terms of this new minimal theory.

Interestingly, while Galton predictably returns to a metaphor here, he does not resort to a quincunx-style argument, in which variability is somehow the result of the compounding of a number of small and independent causes, nor do his representative governments make an appearance. Rather, he takes up one of Darwin's favorite biological phenomena: biogeography. Imagine, Galton writes, that we have two small, well-planted islands, anchored near to one another in the ocean, surrounded by a small number of islets entirely without life. Over time, "seeds from both of the islands will gradually make their way to the islets through the agency of winds, currents, and birds" (Galton, 1889, p. 10). Of course, the vegetation on these islets will bear resemblance to that of the parent-islands. But thanks to slight differences

---

[d] Insofar as this looks like Galton making reference to a potential (and quite interesting) process of inter-particle selection (for "better qualified" particles), this is a suggestion upon which neither Galton nor any of his immediate intellectual descendants follow up.

in dispersal, environmental conditions, and so on, the plants that eventually succeed will differ from each of the parents, while still bearing a broad family resemblance. This explanation—insofar as, to echo Galton's concern from 1877, these various causes are not "petty" influences like those in the quincunx, but rather significant ones—should reopen Galton's worry about our lack of an explanation for the fact that these influences would result in normal distributions, but he doesn't seem to notice that here.

Galton then applies the island metaphor to the problem of latent and patent characters. We know from experience that seeds can lie buried, dormant in the ground for a number of years. Should one of these seeds happen to break loose and find fertile soil on an islet, we could see the reappearance of a plant that had not lived on either of the parent islands in many decades. Hence, the reemergence of latent characters after a number of generations.

Galton has thus, with this metaphor, saved the phenomena—at least in a manner of speaking, given that the connection between the actual processes of particulate character distribution and this rough metaphor are as yet unknown. If something in those real processes works roughly like dispersal toward new islands, then the result will be accurate to our observations. The best we can do at the moment, however, is give a characterization of the various categories into which those actual processes might fall. First, as Darwin had it in the original theory of pangenesis, it might be the case that each element has a sort of chemical affinity, via which it "selects its most suitable immediate neighborhood," being attracted to other particles that produce the same kind of biological feature. Second, there might be a kind of structuring cause arising from the whole mass of elements, on which we see a "more or less general co-ordination of the influences exerted on each element, not only by its immediate neighbors, but by many or most of the others as well." Or, last, the particles might just arrange themselves "by accident or chance, under which name a group of agencies are to be comprehended, diverse in character and alike only in the fact that their influence on the settlement of each particle was not immediately directed towards that end. In philosophical language we say that such agencies are not purposive, or that they are not teleological" (Galton, 1889, p. 19)—a nice invocation of Darwin's sense of accident as opposed to design. But a better account than this broad categorization, or any real understanding of the kind of process that governs the distribution of any real-world traits, for the moment at least, lies beyond the reach of science.

For his part, Galton knows that this is a somewhat rough-and-ready characterization of particulate inheritance, and that we lack the kind of biological detail that could truly both confirm this picture and demonstrate the way in which it might give rise to the statistical patterns to which he will shortly turn. It is apparent, however, that this vagueness is entirely intentional. Having been bitten once by the failure of pangenesis to pan out experimentally, Galton wishes to make no such mistake a second time. "I have largely used metaphor and illustration to explain the facts," he writes, "wishing to avoid entanglements with theory as far as possible, inasmuch as no complete theory of inheritance has yet been propounded that meets with general acceptation" (Galton, 1889, p. 34). As he sums the matter up in the conclusion to the book, "it was not necessary…to embarrass ourselves with any details of theories of heredity beyond the fact, that descent either was particulate or acted as if it were so" (Galton, 1889, p. 193). This glib claim, however, belies the depth of work that Galton has put into the question of particulate inheritance, and as I will argue shortly, the open research question that Galton leaves surrounding the behavior of these elements becomes one of the most significant research programs which the tradition following Galton will take from his efforts.

Of chances and causes

Since it turns out that there are precious few theoretical claims which we can confidently make about the nature of particulate inheritance, what would ground our use of statistics and chance in understanding the results of these hereditary transmission processes? First and foremost, Galton argues, we must keep in mind that we are only using statistical tools with the aim of understanding how the real, underlying processes of particulate inheritance might work in the biological world. There is no statistics to be found here for its own sake. As he writes,

> [The first chapter's] intention has been to show the large part that is always played by chance in the course of hereditary transmission, and to establish the importance of an intelligent use of the laws of chance and of the statistical methods that are based upon them, in expressing the conditions under which heredity acts.
>
> **(Galton, 1889, p. 17)**

Again, it is important to draw a contrast with the view of Hacking, on which chance and statistics are somehow playing an autonomous role in biological explanations. Galton's goal—in the face of difficult barriers to obtaining empirical evidence and primitive statistical methods—is always to determine the characteristics of the fundamental processes in particulate inheritance that could, because they are the "conditions under which heredity acts," give rise to future insight about what Galton here calls heredity in general, the transmission of characters over generational time— none of which is to be identified with these statistical tools.

Galton emphasizes the same point elsewhere, as well. He describes the appearance of normal curves across a number of human characteristics as merely approximate, and notes that "with reasonable precautions we may treat them as if they were wholly so [i.e., precisely normally distributed], in order to obtain approximate results" (Galton, 1889, p. 59). This use of unrefined methods, he analogizes, is rather like that of a logger who estimates the volume of a tree as though it were perfectly straight and exactly cylindrical. Such approximate character is in some sense inescapable. "A parade of great accuracy is foolish," he admonishes us, "because precision is unattainable in biological and social statistics; their results being never strictly constant." What we need, therefore, and what statistics provides us, is "no more than a fairly just and comprehensive method of expressing the way in which each measurable quality is distributed among the members of any group" (Galton, 1889, p. 36). In his conclusion, he writes that statistics "were found eminently suitable for expressing the processes of heredity," in an "exceedingly simple form of approxima-tive accuracy" (Galton, 1889, p. 193). Statistics is not here, it seems, intended to pro-vide replacements for the underlying story that we ought to still strive to complete. While we will return to the question of evolutionary dynamics in the next section, this reading is reinforced by the fact that Galton seems to think that the processes of change in organic systems are essentially independent of these statistical distri-butions. To understand change, we must build new and different mechanical mod-els, adding influences and explanations that lie outside the statistical representations themselves—recall, for instance, the behavior of Galton's modified quincunx which artificially reduces variance. Statistics is about approximation and statics; accuracy and dynamics must follow on after.

Galton's habitual sloppiness in argumentation, however, makes the question of the connection between statistics and the underlying biological processes at issue signifi-cantly more difficult, and it is worth seeing the evidence he offers us for the interpreta-tion contrary to the one I take here. After comparing the statistical parameters of two generations in human stature, Galton writes that the values for mean and variance in the latter generation "are identical with those in Generation I.; so the cause of their statistical similarity is tracked out" (Galton, 1889, p. 117). Setting aside some possible ambiguity in the archaic "tracked out" (is this intended to declare that the causes have really been discovered, or merely that we know how they will in fact have to act, having described the phenomena?), it does seem as though Galton is equating knowledge of statistical parameters with knowledge of causes. In introducing the normal distribution itself, and discussing its having arisen in the context of error in astronomical measure-ment, Galton writes that "Errors, Differences, Deviations, Divergencies, Dispersions, and individual Variations, all spring from the same kind of causes" (Galton, 1889, p. 55)—but without telling us what kind of causes these are. It is possible that he has in mind here a picture of all normal distributions arising through small accidents, in the way that Quetelet's derivation did. He writes just afterward that the law of error can be used "wherever the individual peculiarities are wholly due to the combined influence of a multitude of 'accidents'" (Galton, 1889, p. 55), but this again runs afoul of the nu-merous places in his corpus where he worries about how it is that a normal curve might arise in the absence of such a "petty influence" structure.

We thus have to navigate a somewhat contradictory terrain in order to determine Galton's true position on the relationship between statistical thought and the underly-ing biological world. For my part, the balance of the evidence seems to point in favor of Galton's deep commitment to the necessity of grounding statistical inquiry in the nature of particulate inheritance, constrained as it was by the near-impossibility of obtaining the relevant empirical data and Galton's occasionally sloppy argumenta-tion to the contrary.

Natural selection, or supposed to be

An equally confounding facet of Galton's work concerns his approach to natural selection, which combines a degree of confusion about whether evolution proceeds by gradual steps or large jumps (i.e., gradualism vs. saltationism) with simplified examples that shed little light upon the subject. Galton's primary tool for thinking about natural selection is yet another mechanical model. He asks us to consider an oblong, faceted stone (Fig. 2.4). The statistical stability to which he frequently refers, and which he has taken to be his primary explanatory target (for, at this point, a number of decades), is analogous to the resting of this stone on one of its long faces. The expected behavior of this system when it is perturbed is that, even when slight deviations occur, they are too weak to turn the stone enough to fall onto another face. Stasis is therefore the order of the day. On the other hand, "every now and then the offspring of these deviations do not tend to revert [to the ancestral mean], but possess some small stability of their own" (Galton, 1889, p. 28)—rather like the stone coming to rest on a small face adjacent to the first, in an unstable equilibrium. In such a position, the stone is still likely to return to the previous state, but is also "capable of becoming the origin of a new race with very little assistance on the part of natural selection" (Galton, 1889, p. 28), were it to topple over to a new long face. This appears to be a saltationist model, by which evolution proceeds via fits and starts, new "sports" being able occasionally to push the type to new stable means (for an illuminating discussion of the model, see Gould, 2002, pp. 343–351).

On the other hand, Galton speaks often of natural selection—where it is clear that he means, or intends to mean, a notion just as gradual and incremental as Darwin's—as being entirely compatible with his picture of inheritance. Selection, he writes, "may harmonize with [the action] of pure heredity, and work together with it in such a manner as not to compromise the normal distribution of faculty" (Galton, 1889, p. 119). Unfortunately, the example of selection which he treats statistically is a very peculiar one indeed (and one which, it is strange to say, Galton never seems to have recognized as a special case, since introducing it in the 1877 lecture). "The case," he writes, "that best represents the various possible occurrences [is] that in which the mediocre members of a population are those that are most nearly in harmony with

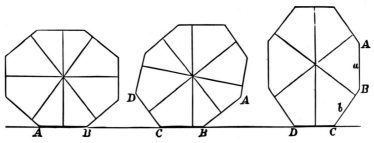

FIG. 2.4

The "faceted stone" which Galton uses as an analogy for the action of natural selection, in a figure taken from *Natural Inheritance* (1889).

Credit: Galton, F., 1889. Natural Inheritance, Macmillan, London, p. 27, figure 1.

Public domain book, available from the Internet Archive at <https://archive.org/details/

galton-francis.-natural-inheritance-1889_202008/page/n3/mode/2up>.

their circumstances" (Galton, 1889, pp. 119–120). That is, he has us consider only the situation in which the selective optimum is already identical with the mean value for the character within the population. In his favored example of height, for instance, he writes that "if their race is closely adapted to their circumstances in respect to stature, the most favoured stature would be identical with the M [mean] of the race" (Galton, 1889, p. 120). A strange account of selection indeed, given that it seems unable to explain selection's ability to change the characteristics of a population.

While we might have thought that such an understanding of selection had to do with the technical limitations of Galton's statistical methods, a bit further on he offers us an argument intended, surprisingly, to justify it. We could imagine, he writes, a case in which natural selection indeed would favor the "indefinite increase of numerous separate faculties, if their improvement could be effected without detriment to the rest" (Galton, 1889, p. 123). But this is in fact impossible. Organisms are systems so thoroughly interconnected that it is impossible to "freely" modify a characteristic without imposing trade-offs against a variety of other characteristics. Since organisms are long-established members of their types, modifying a single variable is liable to simply destroy the harmony between parts, rather than optimize it. Having elaborated an example of different varieties of warships to make his point, he concludes that "evolution may produce an altogether new type of vessel that shall be more efficient than the old one, but when a particular type of vessel has become adapted to its functions through long experience it is not possible to produce a mere variety of its type that shall have increased efficiency in some one particular without detriment to the rest. So it is with animals" (Galton, 1889, p. 124). Again, we have a combination of a resolute defense of natural selection and its importance, but with a conclusion that appears profoundly saltationist.

I am not the only commentator to approach Galton's work on natural selection with some degree of confusion. As Jean Gayon put it, "Galton's theoretical analysis of selection impressed not so much by its rigorous demonstration as by the theoretical possibilities it revealed, and by the wealth of images it introduced" (1998, p. 154). Peter Bowler (2014) has perceptively argued that Galton's ambiguity here was a persistent problem for a variety of biologists who followed in Galton's footsteps. Bowler notes that St. George Mivart and Alfred Russel Wallace both read Galton as an anti-Darwinian saltationist, while T. H. Morgan argued on the contrary that Galton was a Darwinian who simply saw it necessary to consider "sports" as a source of variation. One famous textbook author of the early twentieth century, W. K. Brooks, cited Galton's support for both positions and highlighted their contradictory nature. All in all, Bowler concludes, "it seems that Galton was simply ambiguous on the question of the role played by selection" (Bowler, 2014, p. 274).

Natural selection, therefore, is a particular sore spot within Galton's work, and one that, we will see, drew much engagement in the years after the publication of *Natural Inheritance*. Over the course of the volume, Galton pointed out several other open research problems that seem, in hindsight, historically prescient. He gestured at the existence of "alternative inheritance" (in modern terms, non-blended inheritance which would later be important in the debate over Mendelism; 1889, p. 139). He would offer a clear formulation, but no deep theoretical discussion, of the Law of Ancestral Heredity, which would be taken up by the biometricians as a way in which we might quantify the contribution to an offspring not only of the parents, but also of more distant generations (Galton, 1889, p. 135; this had also appeared in earlier works). He even attempts to lay out some speculative conclusions about the micro-level structure of particulate inheritance, attempting to derive at one point the ratio between latent and patent elements (Galton, 1889, p. 187).

Why Galton?

With this, Galton's major output on questions of heredity more or less came to an end, as he spent the last decade of his life working on public support for eugenics. Indeed, as debates raged over heredity, chance, and statistics in the years that followed, we will see a wide variety of figures with radically different theoretical commitments cite Galton's approval as evidence for their positions. It is high time, therefore, that we return to the paradox with which we began the chapter. As has been made clear in the intervening pages, Galton is a deeply frustrating figure. His arguments often transition to physical models just at the point when their conceptual details become most important, and he is sloppy to the point of being regularly contradictory, leaving in his wake confusions that embroil not only our contemporary understanding, but were problematic for practitioners in Galton's immediate aftermath. What, then, was it that biologists like Pearson and Weldon saw as so trenchant in Galton's thought?

Let's begin by looking at some of the phenomena that led Galton to pursue his investigations in the first place. First, we have the prevalence of reversion to ancestral characteristics. As Galton frequently mentioned, he was often struck by the fact that organisms occasionally appeared not to evolve gradually at all, but rather produced characters that dramatically hearkened back to those of far-distant ancestors. Darwin, in his discussion of pigeons, had been amazed by the same thing—one of the best pieces of evidence he adduces for the claim that all pigeons, however fancy and derived their current breed might be, are descended from a single common stock is the fact that occasionally they produce offspring that look for all the world like garden-variety street pigeons (Darwin, 1859, p. 25). Galton takes this sort of evidence very seriously—so seriously, in fact, that it serves as one of the main pieces of data that he believes his picture of heredity should encompass.

This led to Galton's thinking about the connection between heredity and deep ancestry. Such a connection cannot be straightforward, given that some of the characters implicated in cases of reversion might not have been expressed in any ancestor for many generations. Thus he was pushed directly to the proposal of the distinction between latent and patent characters, as a consequence of his desire to save the phenomena which he found most interesting. In turn, this produces what is, I argue, the first major theme—perhaps "open question" or even "research program" would be more accurate—to persist from Galton's work. Galton was convinced, as were the biologists that followed him, that statistics would be a method through which we could express, understand, and test hypotheses for the way that these properties of latency and patency could be distributed over time. That is, statistics proved to be an excellent approach to considering the nature of particulate inheritance. We will see echoes of this concept throughout the rest of the authors that follow here.

Second, consider the manner in which Galton in fact applies statistics to the underlying causes of inheritance. As we saw, it is fairly simplistic. Galton sees the primary use of statistics to be offering us a way in which to quickly and approximately describe the distributions of characters found in a large population, and to compare their state across multiple populations—most often with the goal of demonstrating that two distributions are, in fact, the same. Regression complicates the story a bit, giving him a way to examine at least one kind of relationship between two different distributions. But this is nonetheless a largely descriptive enterprise. We see a similar approach in his understanding of natural selection. Selection is not seen as an agent of change in populations, but rather as something perhaps more closely akin to the contemporary notion of stabilizing selection, an influence which allows populations to transmit their statistical parameters, more or less without modification, to their descendants.

Galton's picture thus owes perhaps more than might be initially recognized to that of Quetelet. Quetelet's view—despite his occasional protestations to the contrary that he does, indeed, believe in the eventual malleability of the human race—places statistics in the service of a science that is strikingly static (all the more so given that he terms it "social physics"). We can for instance describe, Quetelet argues (and laments!), precisely the number of poisoners that we should expect to arise in every generation. Galton's picture is broadly the same—he gives us practically no way in which to understand the evolving dynamics of a population, having neither a statistically described process of natural selection that could push for change nor any simple statistical techniques (reversion to a different mean, non-normal or asymmetric distributions of characters) that could have led to changes over time.

Galton thus leaves wide open the problem of evolutionary dynamics. For committed Darwinians working—as Pearson, Weldon, and those who came after them were—in an environment where the ability of evolution to ground population change is precisely one of the most bitter loci of biological argument, we cannot merely gesture at questions of dynamics, as Galton did, as special cases to be figured out once the "typical laws" had been fully understood. This is the second open research program left behind by Galton's work—how can we synthesize a statistical approach, previously associated only with relatively static distributions of characters, with an understanding of evolutionary change?

Thus we have Galton's "powerful if vague" impact on the future of the field. The twin questions of ancestry (or particulate inheritance) and evolutionary dynamics would set a baseline for the agenda in the coming years of research in what would become known as the biometrical school. In the absence of solutions to either project, Galton's work in and of itself is of uncertain impact. Had there not been a school which developed around it and proceeded to take up such questions, it is by no means obvious what the legacy of Galton's work on inheritance (setting aside his program in eugenics) would have been on biology at large. At the same time, it is clear, thanks only to his persistent ambiguity, how Galton could preserve his role as an elder statesman of the life sciences, even in an environment as contentious as that which would emerge around the debate between the biometricians and Mendelians. To see how that debate emerged, it is time to turn to one of the seminal works in the biometrical school, one which would, in fact, see the light of day largely due to Galton's own handiwork.

References

Ariew, A., Rice, C., Rohwer, Y., 2015. Autonomous-statistical explanations and natural selection. The British Journal for the Philosophy of Science 66, 635–658. https://doi.org/10.1093/bjps/axt054.

Ariew, A., Rohwer, Y., Rice, C., 2017. Galton, reversion and the quincunx: The rise of statistical explanation. Studies in History and Philosophy of Biological and Biomedical Sciences 66, 63–72. https://doi.org/10.1016/j.shpsc.2017.08.001.

Bowler, P.J., 2014. Francis Galton's saltationism and the ambiguities of selection. Studies in History and Philosophy of Biological and Biomedical Sciences 48B, 272–279. https://doi.org/10.1016/j.shpsc.2014.10.002.

Darwin, C., 1859. On the origin of species, first ed. John Murray, London.

Depew, D.J., Weber, B.H., 1995. Darwinism evolving: systems dynamics and the genealogy of natural selection. Bradford Books, Cambridge, MA.

Galton, F., 1863. The first steps towards the domestication of animals. Spottiswoode & Co., London.

Galton, F., 1865a. Hereditary talent and character [part I]. Macmillan's Magazine 12, 157–166.

Galton, F., 1865b. Hereditary talent and character [part II]. Macmillan's Magazine 12, 318–327.

Galton, F., 1869. Hereditary genius: an inquiry into its laws and consequences, first ed. Macmillan, London.

Galton, F., 1871. Experiments in pangenesis, by breeding from rabbits of a pure variety, into whose circulation blood taken from other varieties had previously been largely transfused. Proceedings of the Royal Society of London 19, 393–410.

Galton, F., 1876. A theory of heredity. Journal of the Anthropological Institute of Great Britain and Ireland 5, 329–348.

Galton, F., 1877a. Typical laws of heredity I. Nature 15, 492–495. https://doi.org/10.1038/015492a0.

Galton, F., 1877b. Typical laws of heredity II. Nature 15, 512–514. https://doi.org/10.1038/015512b0.

Galton, F., 1877c. Typical laws of heredity III. Nature 15, 532–533. https://doi.org/10.1038/015532a0.

Galton, F., 1885. Presidential address, Section H, Anthropology. Report of the British Association for the Advancement of Science 55, 1206–1214.

Galton, F., 1886. Hereditary stature [President's address]. Journal of the Anthropological Institute of Great Britain and Ireland 15, 488–499.

Galton, F., 1889. Natural inheritance. Macmillan, London.

Galton, F., 1892. Hereditary genius: an inquiry into its laws and consequences, second ed. Macmillan, London.

Galton, F., 1908. Memories of my life. Meuthen & Co., London.

Gayon, J., 1998. Darwinism's struggle for survival: heredity and the hypothesis of natural selection. Cambridge University Press, Cambridge.

Gould, S.J., 2002. The structure of evolutionary theory. Harvard University Press, Cambridge, MA.

Hacking, I., 1990. The taming of chance. Cambridge University Press, Cambridge.

Kevles, D.J., 1985. In the name of eugenics: genetics and the uses of human heredity. University of California Press, Berkeley.

Norton, B.J., 1978. Karl Pearson and the Galtonian tradition: studies in the rise of quantitative social biology (Ph.D Dissertation). University of London, London.

Pearson, K., 1889. On the Laws of Inheritance According to Galton.

Pearson, K., 1906. Walter Frank Raphael Weldon. 1860–1906. Biometrika 5, 1–52. https://doi.org/10.1093/biomet/5.1-2.1.

Pearson, K., 1930. The life, letters, and labours of Francis Galton, volume 3A: correlation, personal identification, and eugenics. Cambridge University Press, Cambridge.

Pence, C.H., 2015. The early history of chance in evolution. Studies in History and Philosophy of Science 50, 48–58. https://doi.org/10.1016/j.shpsa.2014.09.006.

Porter, T.M., 1985. The mathematics of society: variation and error in Quetelet's statistics. British Journal for the History of Science 18, 51–69.

Quetelet, A., 1835. Sur l'homme et le développement de ses facultés, ou Essai de physique sociale. Bachelier, Paris.

Quetelet, A., 1846. Lettres sur la théorie des probabilités appliquée aux sciences morales et politiques. M. Hayez, Bruxelles.

Quetelet, A., 1848. Du système social et des lois qui le régissent. Guillaumin, Paris.

Quetelet, A., 1871. Anthropométrie, ou mésure des différentes facultés de l'homme. C. Muquardt, Bruxelles.

Schwartz, J., 2008. In pursuit of the gene: from Darwin to DNA. Harvard University Press, Cambridge, MA.

Stigler, S.M., 1986. The history of statistics: the measurement of uncertainty before 1900. Cambridge, MA, Belknap.

Swinburne, R.G., 1965. Galton's law—Formulation and development. Annals of Science 21, 15–31. https://doi.org/10.1080/00033796500200021.

The only ultimate test of the theory of natural selection: The early years of biometry

I am beginning to find multiplication and division fail to be amusing,
after a time.
Weldon, letter to Galton, 21 January 1890

As we have already seen in appraising the volume's impact at the beginning of the last chapter, a certain invertebrate zoologist named W. F. R. Weldon (he went by Raphael; Fig. 3.1) read Galton's *Natural Inheritance* shortly after it was published in 1889. Weldon was an accomplished, if quite young, scientist, who had made a reputation for himself as adept both at field-work, in the zoological stations at Naples and Plymouth, and as an engaging instructor to his students at Cambridge, where he had found work lecturing with the help of Adam Sedgwick. Weldon is immediately drawn to the possibility of expressing variation in natural populations using the tools of statistical distribution as they were developed by Galton—Pearson, in his obituary for Weldon (1906), claims that the paper which would result (Weldon, 1890a) is the first-ever application of statistical methods to animals other than man, and I can find no evidence to the contrary. Why did Weldon turn to statistical methods? What did he hope to bring from them to the study of evolution by natural selection? The answer is a long and circuitous one, and it will be the topic of both this chapter and the next. But by the time we traverse the short time from 1890 to 1906, we will find the science of evolution looks entirely different from that of Darwin or even Galton. The biometrical school will steadily refine a set of desiderata for a chancy approach to natural selection over these two decades, a way of theorizing about evolving populations grounded in a philosophy of science that made a gradualist, mathematized, selection-first view of the living world possible.

The period begins with Weldon drafting an article correlating a vast number of measurements of organs in a particular shrimp (then *Crangon vulgaris*, now *Crangon crangon*, familiar to residents of Belgium and the Netherlands as the delicious *crevettes grises* or *gewone garnaal*; Fig. 3.2) and sending it off to the *Proceedings of the Royal Society*. Galton himself is selected as his peer reviewer. No details of Galton's original critique survive, but we do have the first letter that Weldon writes in response to Galton, dated the 7th of January 1890. It would seem that Galton was not impressed, and Weldon is, evidently, quite worried:

The Rise of Chance in Evolutionary Theory. https://doi.org/10.1016/B978-0-323-91291-4.00009-1

FIG. 3.1

W. F. R. Weldon.

Credit: Pearson, K., 1906. Walter Frank Raphael Weldon. 1860–1906. Biometrika. 5 (1/2), 1–52. p. 1–1 (overleaf). Oxford University Press. Public domain image.

FIG. 3.2

The shrimp *Crangon vulgaris* (now known as *Crangon crangon*).

Credit: Photo by Hans Hillewaert, licensed under CC-BY-SA 4.0, available on Wikimedia Commons at https://commons.wikimedia.org/wiki/File:Crangon_crangon.jpg.

My dear Sir,

I have had a long dose of Influenza in Dresden, during which I have seen no let-ters: so that your condemnation of my paper to the Royal Society has only just reached me.

You will easily understand that I am anxious to learn the nature of my triple mis-conception as soon as possible: so that if you can allow me to call upon you sometime within the next day or two I shall be very grateful.

Let me thank you for the very kind way in which you have tried to make the con-sequences of my blunders as easy for me as may be.

   Yours very truly,

   W. F. R. Weldon   **(Weldon, 1890b)**

Over the next few months of ensuing correspondence, the reason for Weldon's diffi-culties will become readily apparent. Weldon has taken on, as his first-ever statistical problem, a scenario that (particularly by comparison with the prior work of Galton) is impressively difficult to manage. First, one of his measured characters of interest in one of his populations turns out to produce a *non*-normal distribution—an asymmet-ric curve, a case that, as we saw, was not considered by Galton at all in the analyses of *Natural Inheritance*. Second, Weldon is immediately interested in the possibility that this kind of asymmetric distribution could be connected with the action of natu-ral selection. Galton writes to Weldon, apparently providing a statistical method that could approach the problem. (Weldon, sadly, kept only very little of his incoming correspondence; we are often forced to reconstruct these dialogs from only Weldon's side of the conversation.) But Weldon is skeptical, and spends time working with Donald MacAlister, a Cambridge mathematician, physician and later Chancellor of the University of Glasgow, with the aim of improving his statistical abilities. "I…am wrestling with the effect of selection on the Normal curve," he writes to Galton, now a few weeks after their initial contact. "Your asymmetrical result seems so hard to understand. In fact I cannot, without trial, understand how anything but the original curve can be reproduced" (Weldon, 1890c). Galton's efforts, such as they were, to derive the non-normal distribution that Weldon was seeing in these shrimp seem not to have satisfied Weldon, producing (once again) only stasis.

   As it turns out, for the purposes of these data on shrimp, the question of asym-metric curves and natural selection would be dropped. Weldon returns to the field, adds several hundred more observations, and the apparent asymmetry that he had seen in the curves simply vanishes. "I suppose then that even I cannot doubt anymore the complete normality of the curves of frequency of these variations," he writes to Galton on February 16, "and these curves must be taken as a complete demon-stration of the truth of your remarks on Natural Selection (pp. 119–121 of Natural Inheritance)" (Weldon, 1890d). This would, in turn, be the position of his final article as it appeared in the *Proceedings* (the paper would be communicated around a month later, on March 20, and published on April 17). He writes there,

In his recent work on heredity, Mr. Galton predicted that selection would not have the effect of altering the law which expresses the frequency of occurrence of deviations from the average: so that he expected the frequency, with which

deviations from the average size of an organ occurred, to obey the law of error in all cases, whether the animals observed were under the action of natural selection or not. The results of the observations here described are such as to fully justify Mr. Galton's prediction.

(Weldon, 1890a, p. 446)

Weldon, who had clearly been expecting from his shrimp curves some statistical results that at least in some way differed from Galton's approach ("they throw light upon very little!" he laments at the end of a letter to Galton; Weldon, 1890d, fol. 3r), has no choice but to offer his data as further confirmation of Galton's view.

So much for this opening vignette. But several morals are already worth drawing here, because they indicate to us the direction of Weldon's thought before he engaged substantively with either Galton or Pearson. To continue with the unresolved questions left behind by the narrative of the last chapter, we see that Weldon's very first interest in the collection of data is centered around demonstrating the action of natural selection in real-world populations—with "the effect of selection on the Normal curve," as he put it above. Pearson is quite right when, in his obituary for Weldon, he writes that "the credit…of seeing *a priori* the bearing of his results on the great problems of evolution, must be given to Weldon" (Pearson, 1906, p. 17).

This comes into even clearer focus if we move just a touch further into the February letter from Weldon to Galton already discussed above. Just after admitting that he thinks Galton has things right after all (i.e., that normality will remain in a population undergoing natural selection), Weldon continues:

But there is a feature about this which troubles me: namely, that I see no way of demonstrating *that the curve is a result of Natural Selection. You have a brood of newly born shrimps with [a given variability or standard deviation]. These give rise to an adult population whose [standard deviation is smaller]. This may conceivably be done either by the [coincidental] modification of each individual in such a direction that its deviation from the median is reduced, or it may be done by Selection, as shown in your book. Can you suggest any statistical test which will determine* which *influence is acting in any given case?*

(Weldon, 1890d)

This is, of course, quite the statistical challenge, as Weldon recognizes (calling it a "fine old problem which has been suggested by many people"). Galton proposes an experiment—something to do with raising a number of shrimp from their larval stage and tracking them through their later lives, though the details are hard to reconstruct from only Weldon's letter in response—which Weldon rejects out of hand. The set of causal influences on such larva, including those as diverse as algae, protozoa, other larva, small adult crustaceans, fish, and even environmental conditions such as rain and salinity, is just too large. More importantly from a conceptual perspective, they encompass both selective and non-selective causes of destruction. While it is easy to divide such influences conceptually, Weldon laments that "I am afraid no experiments on animals in captivity will enable me to estimate the relative efficiency of these two classes of forces" (Weldon, 1890e)—the distance between any manipulations that might be made in captivity and the utter pandemonium of such forces as they impinge upon larvae in the wild would simply be too great.

But to move any farther into Weldon's work on the nature of evolution and selection would be to outrun our story here. We must briefly double back.

Pearson before biometry

In 1884, 6 years prior to Weldon's publication, a young mathematician named Karl Pearson was appointed to the chair of Applied Mathematics and Mechanics at the relatively new institution of University College, London, where he taught mathematics primarily to students of engineering (Pearson, 1936, pp. 206–207). Pearson was already, at the age of 27, a dramatic and interesting character, to whose life I cannot begin to do justice here. As Theodore Porter details in his masterful biography covering the first half of Pearson's life, Pearson had by this point "published on the cultural history of Germany in the Reformation period and tried his hand at a nineteenth-century passion play. He was then beginning to think systematically about 'the woman's question'" (Porter, 2004, p. 14). He had written a version of his life as a romantic autobiography entitled *The New Werther* (after Goethe's classic), and changed the spelling of his name from Carl to Karl 4 years before his professorial appointment, out of a feeling of kinship with Germanic culture and philosophy. Porter is quite right when he muses that Pearson's life is tailor-made for the plot of a novel.

For our purposes, two main threads will suffice to situate Pearson's thought prior to the beginning of his collaboration with Weldon, which would define the bulk of his work from 1893 until 1906, and which will be my real target in this chapter and the next. First is his general outlook on the philosophy of science. Both in his day and in our own, he is best known for *The Grammar of Science*, an 1892 *opus magnum* (with a second edition in 1900, about which more later, and a third, partial edition in 1911) that was considered required reading by a vast number of scientists in the late-nineteenth and early-twentieth centuries (including the young Einstein, who chose it as a topic for a reading group he assembled in Bern). But we need not look even this far forward to see Pearson's views in a nearly mature form. The year after he began work at University College, Pearson published his completion of William Kingdon Clifford's *Common Sense of the Exact Sciences*, which Clifford had left unfinished upon his death (Clifford and Pearson, 1885). A volume in what we would today term the philosophy of physics, Clifford had hoped to offer a unified account of mathematics, space, position, motion, force, and mass. A brief exposition of Pearson's contribution there can help us see the extent to which his philosophical commitments had already begun to crystallize.

As he writes in the introduction, having needed to significantly rework Clifford's notes for the book's chapter on motion, he found that Clifford had never published on the nature of physical law. "I have accordingly expressed," he writes, "although with great hesitation, my own views on the subject; these may be concisely described as a strong desire to see the terms matter and force, together with the ideas associated with them, entirely removed from scientific terminology" (Clifford and Pearson, 1885, pp. viii–ix), replaced instead by dynamical descriptions summarized in mathematical formalism. Even such a brief summary bears clear resemblance to a Machian-positivist understanding of the laws of physics as mathematical shorthand for describing empirical experience, a resemblance which Pearson is quite happy to underline. "I should hardly have ventured to put forward these views had I not recently discovered that they have (allowing for certain minor differences) the weighty authority of Professor Mach, of Prag" (Clifford and Pearson, 1885, p. ix). He then footnotes a citation to Mach's *Die Mechanik in ihrer Entwickelung* (translated as Mach, 1919). (The admiration was mutual: Mach would write to Pearson a dozen years later, in 1897, that "it has pleased me very much to have gotten to know a man with whom I can so truly agree" (Thiele, 1969).) The full expression of this view—given on the very last page of the volume—makes the point even clearer:

> *The custom of basing our ideas of motion on these terms 'matter' and 'force' has too often led to obscurity, not only in mathematical, but in philosophical reasoning. We do not know* why *the presence of one body tends to change the velocity of another; to say that it arises from the force resident in the first body acting upon the matter of the moving body is only to slur over our ignorance.*
> **(Clifford and Pearson, 1885, p. 271)**

Pearson's positivism—a trend noted by a number of authors throughout the history and philosophy of science literature (e.g., Pence, 2011; Radick, 2005; Sloan, 2000)—is thus a foundational element of his approach to and understanding of the sciences. The basic ideas that he developed in his earliest philosophical work, including that scientific observation is about systematic, mathematical description and that the inference to "metaphysical" factors or influences (like forces, in the example here) is to be scrupulously avoided so as not to lead science astray, are carried forward from these very first writings on the nature of the scientific method.

The other thread which can situate Pearson's early perspective comes from his initial engagement with statistical biology. Already extremely interested in the intersection of feminist and eugenic thought, Pearson had founded the "Men's and Women's Club" in 1885, as a place to foster frank discussions about relations between the genders, eugenics, procreation, sex, and family structure (Porter, 2004, p. 125). In the middle of a series of lectures for the group on the topic of heredity that extended over the course of late 1888 and early 1889, Galton's *Natural Inheritance* was published. Pearson read it, and was interested enough by it to suggest that the club's upcoming meeting (which was to have been about, Pearson notes, the much more interesting topic of women's education) should be devoted to a discussion of Galton's work. Thus, on March 11, 1889—some 10 months before Weldon's first letter to Galton—Pearson delivers a lecture to the club entitled "On the Laws of Inheritance According to Galton."

We can discover several things in Pearson's early reflections. First of all, he seizes on Galton's penchant for describing the stasis of the natural world:

Briefly his book may be described as a theory concerning the most probable amount of any quality which an individual must inherit in order that the average amount of that quality in the multitude may remain the same for long generations. Briefly it is the theory of what is the probable variability in the individual which is consistent with the preservation of mediocrity by the multitude.

(Pearson, 1889, fol. 3)

Galton therefore endeavors to present us with a theory of long-term, population-level stasis. Pearson dedicates much space to Galton's sources of biometric data (connecting with his emphasis on eugenics), and he introduces the "laws of chance" not in the way that Galton does, by reference to the work of Quetelet, but via the much older tradition of the treatment of errors in large numbers of measurements of a physical constant (about which more in the next chapter, as this framing will later be very important to Weldon).

Second, and most importantly, however, is Pearson's general pessimism about Galton's entire project—a pessimism that is shocking given our knowledge of Pearson's later role in the project of statistical biology as a whole. The passage merits quoting at length:

Personally, I ought to say that there is, in my own opinion, considerable danger in applying the methods of exact science to problems in descriptive sciences, whether they be problems of heredity or of political economy; the grace and logical accuracy of the mathematical processes are apt to so fascinate the descriptive scientist that he seeks for sociological hypotheses which fit his mathematical reasoning, and this without first ascertaining whether the basis of his hypotheses is broad as that human life to which the theory is to be applied. I write therefore as a very partial sympathiser with Galton's methods.

(Pearson, 1889, fols. 1–2)

The early Pearson—immersed as he was, at the time, in writing on the physical sciences—sees the introduction of statistical methodology into the life and social sciences as, above all, a reason for caution, lest we be carried away by the (perhaps merely) apparent clarity of mathematical conclusions.

What, then, caused him to change his mind? In short, it was the arrival of Weldon. In December 1890 (8 months after we left him in his correspondence with Galton), Weldon, now a newly minted Fellow of the Royal Society, takes up a professorship of his own at University College. Their first few years as colleagues are mired in some degree of mystery; neither the archival materials for Pearson, nor those for Weldon, are particularly illuminating. In his obituary for Weldon, Pearson writes that this time, on the one hand, encompassed "strenuous years in calculating" (Pearson, 1906, p. 18), but on the other hand, that both of Weldon's early papers on shrimp (those of 1890 and a follow-up published in 1892) were written before their collaboration had truly begun. In his memorial for his father, Egon Pearson writes that Karl was still occupied in these years with his work in physics (particularly his history of the theory of elasticity, which was published in 1890) and the fashioning of the first edition of the *Grammar of Science*, which does not include any notable evolutionary thought (Pearson, 1936, p. 211). Concrete evidence of Pearson's steady conversion to work on a mathematical approach to evolutionary theory is, therefore, rather slim. We do know, however, that this period was marked by extensive discussion between the two men. "Weldon and the present author," Pearson writes, "both lectured from 1 to 2, and the lunch table, between 12 and 1, was the scene of many a friendly battle, the time when problems were suggested, solutions brought, and even worked out on the back of the *menu* or by aid of pellets of bread" (Pearson, 1906, p. 18).

We must look ahead a bit to see the genuine fruits of these early labors. Weldon's 1893 paper, combined with Pearson's first entry in his series of articles entitled "Contributions to the Mathematical Theory of Evolution" (a series that would eventually number in the dozens), which was read in late 1893 and published in 1894, form the first real results of the biometrical school. It is to this work of Weldon's—which, as Pearson writes, "biometricians will always regard as a classic of their subject" (Pearson, 1906, p. 19)—we now turn.

The collaboration in earnest: The crab papers

As Pearson tells it (1906, p. 18), Weldon spent his Easter vacation in 1892 traveling with his wife, Florence—who was his constant companion, partner, and assistant in all matters experimental, calculational, and otherwise—to biological stations at Malta and Naples, to measure the crab *Carcinus mœnas* (now *Carcinus maenas*), a common (and, today, highly invasive) shore crab. His initial goal was to perform exactly the same kind of routine measurements on the crabs as he had done on the shrimp—in fact, in spite of a fairly detailed correspondence with Galton during this period, in which he discusses several experiments to come, he makes almost no mention of the crabs as a system of interest until after the data have been collected.

Finally, however, things begin to come together. On the 27th of November 1892, he writes letters to both Galton and Pearson. We can do no better than Weldon's own account of the process in his letter to Pearson:

> *In the last few evenings, I have wrestled with a double humped curve, and have overthrown it. [...] It represents 1000 measures of an organ in Naples crabs. Body length was taken as [a common unit of] 1000. [...] Now the resulting curve was clearly asymmetrical. I therefore took a table of the Probability Curve, and drew curves whose sum seemed to fit the observations.*

(Weldon, 1892a, fols. 1r–1v)

"If you scoff at this," Weldon writes in closing the letter, "I shall never forgive you" (Weldon, 1892a, fol. 2r). Sadly, despite Weldon having sent two copies of the same diagram, neither Galton's nor Pearson's copies are preserved. The basic idea, however, can be seen from the graph that would be published in the final version of the paper the following year (Fig. 3.3). Weldon has measured a vast number of morphological characters of his crabs (Fig. 3.4), and for precisely one of them, which he calls "frontal breadth" (the distance between points C and D in Fig. 3.4, across the front of the

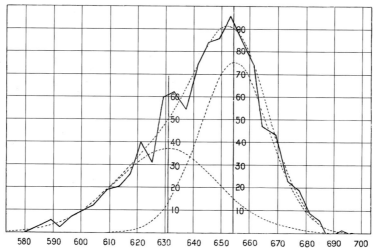

FIG. 3.3

Weldon's asymmetric graph of frontal breadths in the Naples specimens of *Carcinus mœnas*.

Credit: Weldon, W. F. R., 1893. On Certain Correlated Variations in Carcinus mœnas. Proc. R. Soc. 54, 324, fig. 3. Public domain figure, retouched, originally published by The Royal Society.

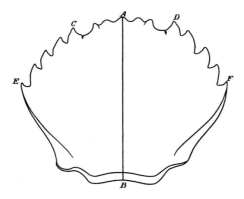

FIG. 3.4

The carapace of a *Carcinus mœnas*, with frontal breadth (the distance *CD*) indicated.

Credit: Weldon, W. F. R., 1893. On Certain Correlated Variations in Carcinus mœnas, Proc. R. Soc. 54, 319, fig. 1. Public domain figure, retouched, originally published by The Royal Society.

crab; Pearson jokingly refers to these as "forehead" measurements), he has detected a clearly asymmetric curve. Unlike in the case of his shrimp, however, no amount of further data collection, more precise measurement, or discarding of malformed outliers seems to make the asymmetry disappear. We may actually have here the non-Galtonian character which Weldon has been hoping to find for several years.

But what does the existence of the asymmetry mean, and how are we to analyze it? Weldon's effort is simple enough. He has—purely by hand-drawn trial-and-error using tables of normal curves—constructed two curves which sum to the overall, asymmetric result which he has seen in nature (these eventually become the two smaller dashed curves in Fig. 3.3, which sum to the larger dashed curve that approximates the solid-line empirical observations). Notably, this is not in the slightest an important moment in the history of statistics. Weldon's graph here is entirely approximate; a proper statistical method for factoring a compound, asymmetric curve into multiple normal curves would only be developed by Pearson in 1893 to generate the published, final version of the graph shown in Fig. 3.3 (presented in abstract form immediately following Weldon's paper; Pearson, 1893). Pearson would publish the complete details of the method of analysis in the *Transactions* in the following year (Pearson, 1894). But it is a pivotal moment for the development of chance and statistics in evolutionary theory. Immediately, Weldon sees the possible significance of these data—as he puts in the opening of his letter to Galton, "it may arise from observation of a 'sport'." As the curve appears to consist of two separate, normally distributed groups, "therefore, either Naples is the meeting point of two distinct races of crabs, or a 'sport' is in process of establishment" (Weldon, 1892b, fols. 1r, 2r).

This is exactly the kind of data for which Weldon has been searching—and it is the first glimpse in the history of biology of the possibility of detecting evolution in action, of seeing natural selection in operation as a process acting to modify populations, represented as statistical distributions of character traits. If the overall population of crabs consists of a mixture of two sub-populations, each of which in turn behave in the expected (Galtonian) manner, we then have two competing hypotheses to test. On the one hand, the crabs might simply have been sampled from a region in which two sub-species happen to overlap. But on the other hand—and more tantalizingly, for Weldon—we might actually have an instance where a population is being split apart by selection in favor of two different character values. In that case, we might be able to demonstrate the action of selection in the wild using a combination of statistical and experimental techniques.

The first paper that Weldon writes, however, does not yet forcefully make the argument for the action of natural selection (Weldon, 1893a). His concern here is almost entirely descriptive. After presenting the graph in Fig. 3.3, he merely writes that "we may, therefore, assume that the female *Carcinus mœnas* is slightly dimorphic in Naples with respect to its frontal breadth; and that the individuals belonging to the two types are distributed in the proportion of nearly two to three" (Weldon, 1893a, p. 324). He then turns to correlational studies between organs, attempting to see if there is any systematic way in which deviations from the mean in one part are accompanied by differences from the mean in other parts.

In the conclusion of the paper, however, he cannot help but offer something of a manifesto for future work. He writes:

> It cannot be too strongly urged that the problem of animal evolution is essentially
> a statistical problem: that before we can properly estimate the changes at present

going on in a race or species we must know accurately (a) the percentage of animals which exhibit a given amount of abnormality with regard to a particular character; (b) the degree of abnormality of other organs which accompanies a given abnormality of one; (c) the difference between the death rate per cent. in animals of different degrees of abnormality with respect to any organ; (d) the abnormality of offspring in terms of the abnormality of parents, and vice versâ.

(Weldon, 1893a, p. 329)

At the risk of whiggishness, we might today describe Weldon's four desired pieces of data as follows: the distributions of characters, the correlations between trait distributions, the fitnesses of different trait values, and the heritability of differences in those traits. In short—an understanding of the statistical nature of selective change that should seem deeply familiar to anyone today working in evolutionary biology. He concludes the paper:

These are all questions of arithmetic; and when we know the numerical answers to these questions for a number of species we shall know the direction and the rate of change in these species at the present day—a knowledge which is the only legitimate basis for speculations as to their past history and future fate.

(Weldon, 1893a, p. 329)

It is important to underline just how revolutionary such a statement was in 1893. Recall that Galton had, strangely, abstained entirely from discussing long-term evolutionary change, and given us no functional account of natural selection. In Weldon's two papers on shrimp that predated this paper, he writes in an entirely Galtonian and, moreover, completely morphological frame. The equivalent call to action from the end of his second paper on shrimp, read just a year earlier, mentions only the ability to obtain a "new kind of knowledge of the physiological connexion between the various organs of animals," along with the "functional correlations between various organs which have led to the establishment of the great sub-divisions of the animal kingdom" (Weldon, 1892c, p. 11). Weldon has thus moved, in the space of around a year and a half, from a Galtonian picture of statistics as a tool to understand morphology to an astonishingly original view, not only that natural selection is a problem which ought to be approached statistically, but even that statistical data are the *only* way in which we might be able to make solid inferences about the action of selection in the wild.

Let's unpack a bit more what this position seems to entail at this point for Weldon. First, it seems that he's already committed to the idea that, at least in the long term, evolution simply is a matter of changes in distributions of traits within populations. Determining evolution's "direction and rate of change" is just a problem of being able to properly classify and interpret those distributions, and knowing the ways in which, on the one hand, having a trait might directly lead to changes in fitness, and on the other, having other traits that are correlated with the trait of interest might affect the latter's dynamics over time. Interestingly—and we see the same point borne out in Weldon's correspondence, particularly as he discusses the construction of various experimental setups with Galton (Weldon, 1890e)—Weldon seems to have a picture of natural selection as a negative force, one that might cause certain values of a trait to disappear by increasing the death-rate of organisms which bear it. Stephen Jay Gould frequently, and poetically, referred to this view of natural selection as being "the headsman for the unfit" (Gould, 1971, p. 249), and a prominent cause of objections to natural selection—for a focus on selection as a way in which to cull unfit organisms calls our attention, in turn, to the (apparently missing) process by which novel, fit organisms might come into being. We will see this problem return for Weldon in the next chapter, as he responds to the rediscovery of Mendel's results by turning in great depth to the study of the material basis of heredity.

At the very least, Weldon took his own manifesto to heart, and before the 1893 paper was even published, was already discussing a new round of data collection which he might undertake with Galton (Weldon, 1893b). The next target is clear enough: if the asymmetric crab data are to really be a demonstration of natural selection in the wild, we must understand the system at issue well enough to rule out competing hypotheses. Unfortunately for Weldon, institutional concerns would intervene. We must pause to consider the context in which the following several years of Weldon and Pearson's work would occur before we can properly appreciate Weldon's further pursuit of natural selection.

Developing a controversy: Weldon and Bateson

Historians and philosophers of biology already familiar with Weldon, or even with general histories of the time period like that of William Provine (1971), will be surprised to have read so much of a chapter about W. F. R. Weldon and Karl Pearson without encountering even the slightest mention of the controversy between the "biometricians" and "Mendelians," classically taken to be the key issue of these several decades in the early history of genetics. In part, this is an intentional choice on my part—Bateson and the early Mendelians had so little grasp of statistical methods or the nature and role of chance in evolutionary theory that they simply have nothing to contribute to the narrative that I am pursuing here, and they will thus feature as nothing more than minor characters. But this debate, perhaps one of the fiercest in the history of the biological sciences, would become so all-consuming that it would drastically alter the course of Weldon's research and professional career; we thus have no choice but to engage with it to at least some degree.

William Bateson, one of the founding fathers of the new science of genetics (the person, in fact, to have coined its name), began his career as a devoted student of morphology. Having become profoundly excited by biology, in no small part due to the enthusiasm of Weldon, who was a year ahead of him in studies at Cambridge, Bateson spent a number of years in traditional morphological study, making, among others, a notable voyage to observe the marine life in the shallow seas of Kazakhstan (Cock and Forsdyke, 2008, pp. 20–23). The major product of his early career was a work entitled *Materials for the Study of Variation*, published in 1894, a compendium not only of his own morphological data collected during his extensive travels, but an impressively detailed survey of the entire state of the art. Understanding morphological variation, he writes, "is, I submit, the first duty of the naturalist. […] Whatever be our views of Descent, Variation is the common basis of them all. As the first step towards the systematic study of Variation we need a compact catalogue of the known facts, a list which shall contain as far as possible all cases of Variation observed" (Bateson, 1894, p. vi). Of course, no work could pretend to completeness with respect to such a mission statement, but the project would offer our best hope of understanding the true nature of evolution and the structure of the organic world.

Bateson and Weldon were, however, set up for conflict from the very first. The subtitle of Bateson's *Materials* is *Treated with Especial Regard to Discontinuity in the Origin of Species*. Bateson was animated from his earliest research by a belief that the variation responsible for giving rise to new species must, *contra* Darwin, Galton (perhaps), Pearson, and Weldon, be fundamentally discontinuous, that is, large enough to produce the gap between species at a single stroke. It is difficult to imagine a passage more at odds with Weldon's view of the world than the following:

> [T]he metaphor of Heredity, through an almost inevitable confusion of thought, suggests the idea that the actual body and constitution of the parent are thus in some way handed on. No one perhaps would now state the facts in this way, but something very like this material view of Descent was indeed actually developed into Darwin's Theory of Pangenesis. From this suggestion that the body of the parent is in some sort remodelled into that of the offspring, a whole series of errors are derived. Chief among these is the assumption that Variation must necessarily be a continuous process; for with the body of the parent to start from, it is hard to conceive the occurrence of discontinuous change.
>
> **(Bateson, 1894, p. 75)**

Note the depths of difference with Weldon's philosophical and empirical outlook that Bateson has packed into a single, short paragraph. The very use of the word "heredity," Bateson argues, is misguided; we ought to neither think that there is direct transmission of some sort of material from parents to offspring, nor even that the body plan of the parent is really the starting point from which that of the offspring varies. To do so denies us the possibility of understanding discontinuous variation, which is how we might hope to really understand the development of new species. Thus, the entirety of the Galtonian-statistical picture, the approach to inheritance not only from parents, but from more distant ancestors ("it would probably help the science of Biology if the word 'Reversion' and the ideas which it denotes, were wholly dropped," he says later; Bateson, 1894, p. 78), the commitment to gradualist change in the Darwinian mold—all of these run contrary to Bateson's worldview.

When Weldon was called upon to review Bateson's book for *Nature* the same year it was published, these differences came to the fore. Weldon begins by heaping praise on Bateson's efforts at systematically presenting the data on variation. "The whole work must be carefully read by any serious student," he writes, "and there can be no question of its great and permanent value, as a contribution to our knowledge of a particular class of variations, and as a stimulus to further work in a department of knowledge which is too much neglected" (Weldon, 1894, p. 25). He then, however, turns to a frontal assault on Bateson's commitment to discontinuity. Why, according to Bateson, should we believe that the variation that produces species is, in fact, discontinuous? Because species are discontinuous, while their environments (at least in terms of physical parameters like temperature, altitude, and so forth) are broadly continuous. Discontinuity must, then, arise through the process of variation. Weldon is not convinced. Setting aside the fact that many environmental parameters are in fact not continuous, Weldon points to an obvious source of discontinuity in the "environment"—interactions with other organisms, a possible cause which Bateson has only dismissed in a cursory footnote. Further, in cases where Bateson has amassed much data of discontinuous variation (his favorites being cases of whole-unit, integer variation like vertebrae or teeth), he has neglected the biases induced by preservation. Extremely small teeth or vertebra are unlikely to be preserved in specimens, and thus without extremely careful study (down to the cellular level, Weldon writes) of samples taken expressly for the purpose, we have no reason to think that the only options are either an entire tooth or no tooth at all. Carefully collected, large-sample-size statistical data would have done better. As he closes the review, "a careful histological account of the jaws of five hundred dogs would have done more to show the least possible size of a tooth in dogs than all the information so painfully collected. And so in many other cases" (Weldon, 1894, p. 25).

Further fuel for the controversy came from unusual quarters. A bitter dispute arose in the correspondence to *Nature* over some examples of cultivated plants in the genus *Cineraria* which had been discussed by W. T. Thiselton-Dyer (1895; see also Cock and Forsdyke, 2008, pp. 150–154). This controversy, in turn, was the result of the presentation of a paper of Weldon's—to which we shall soon turn—which had been developed under the auspices of a Royal Society "Committee for Conducting Statistical Inquiries into the Measurable Characteristics of Plants and Animals," which Weldon had formed in partnership with Galton, Darwin's son Francis, and a few other well-known biologists at the end of 1892. This committee, Weldon had hoped, could serve as a way to centralize statistical work on the measurement of variability in organisms. Bateson, however, saw it as a threat, and set about attacking its work in earnest shortly after the dust-up over the *Cineraria*. The whole episode would end with Bateson being appointed to the committee as a conciliatory gesture, followed by his stacking the committee with his own allies in 1897, and the resignation of the biometricians *en masse* 2 years later, in 1899. In short, it amplified and focused the already deep and profound animus which existed between Bateson's group and that of Pearson and Weldon.

This debate, however, will hardly feature in my narrative. After all, ours is a story about the introduction of chance, probability, and statistics into our picture of the evolving world, and Bateson and his group never really understood those methods well enough to be able to inveigh against them in a serious way. Their critiques remain, rather, largely at the level of a general opposition to the use of mathematical methods in the first place. Furthermore, as we shall see throughout the rest of the book, the historiographical tendency to reduce the contributions of all the figures in this time period—biometrician and Batesonian (later, Mendelian) alike—to simply a series of contributions to a debate between the two groups is uncomfortably reductive, and deprives us of the ability to deeply investigate the work which occurred within this period on its own merits.

Natural selection without (and then with) adaptation

It is, therefore, time that we return to Weldon's efforts to show the action of natural selection in the wild. When we last left him, Weldon had just published his first paper on crabs—demonstrating an asymmetric distribution in a particular feature of his population of Naples shore crabs and launching a broad manifesto in favor of the statistical study of evolution, but lacking, as of yet, the data required to attribute this change to selection as against any number of other competing causal mechanisms. The recapitulation of the problem offered by Jean Gayon is instructive. "If one wants to rigorously demonstrate a process of natural selection," he writes, "one must manage to measure a selective death rate for variations affecting a given character." But since we cannot track the very same singular individuals in a token population as they undergo the effects of natural selection, we must instead

> compare the variability of the population at two stages of ontogenetic development. A reduction in variability, or a modification of the mean character, are serious indications of a process of selection. But nothing prohibits, a priori, thinking that such phenomena can be explained by the "law of growth" of the character. Therefore, the rigorous demonstration of the fact of selection supposes that the law of growth of a character is known.
>
> **(Gayon, 1992, p. 206)**

Such was the problem with which Weldon was grappling when, in late November 1894 (published in the *Proceedings* the following February), Weldon wrote his recounting of the next 7000 female crabs he collected from Plymouth Sound. Recall that in the 1892–93 data, the crabs at Plymouth had in fact been symmetrically distributed. But now Weldon set out to observe not just adult crabs, but crabs ranging in size from 7 to 14 mm. This, he hoped, would give him the data that he required to test the hypothesis that the crabs were undergoing selection against the alternative explanation that they were, in fact, simply growing in such a way as to present a suggestive statistical change in the distribution for some particular character.

This is quite the methodological challenge. Weldon first requires that the data be normalized for the significant change in frontal breadth that occurs naturally as the crabs grow. As it turns out, after Weldon normalizes the frontal breadth values based on the size of the overall crab (i.e., expressing frontal breadths in "units" of crab size), the variation is, at every age, roughly normal. The standard deviation, however, changes—it begins a bit smaller in young crabs, then increases in adolescents, then decreases a bit in the adults. Even here, though, it is impossible to rule out the hypothesis that "the deviation of 'abnormal' young may in each individual case first attain a maximum and then diminish with advancing age" (Weldon, 1895a, p. 368): that is, the hypothesis that this change in standard deviation is simply the way in which the expected pattern of crab growth usually unfolds. But what we can do with the data here, however, is determine whether or not they are consistent with the claim that "the diminution in the frequency of individuals of given deviation is due to a selective destruction" (Weldon, 1895a, p. 368); that is, we can look to see if the death-rate is a function of frontal breadth, and hence demonstrate that the crabs are under the influence of stabilizing selection as opposed to merely indiscriminate destruction. And, as it turns out, this destruction is indeed a function of the particular deviation that the crabs have—if the destruction here is the result of some external cause, that cause is acting selectively.

To be sure, the data remain crude, which limits the extent to which the precise values that Weldon has determined for the influence of selection can be taken seriously. He writes that

It will, of course, be understood that little trust can be placed in the absolute numerical results which are here put forward; the point which seems worthy of confidence, and which if it be indeed a reality is of very great importance, is the form of the result. For by purely statistical methods, without making any assumption as to the functional importance of the frontal breath, the time of life at which natural selection must be assumed to act, if it acts at all, has been determined, and the selective death-rate has been exhibited as a function of the abnormality...

(Weldon, 1895a, p. 371)

Here, then, we have the carrying out of at least part of the manifesto for future work from Weldon's earlier paper—the determination of statistical parameter (c) from his list, what we today might call the trait fitnesses corresponding with various values of frontal breadth. As he writes in concluding the paper, the current state of physiological study for evolutionary purposes is rudimentary enough that such a statistical methodology forms a clear and welcome innovation. "The advantage," he argues, "of eliminating from the problem of evolution ideas which must often, from the nature of the case, rest chiefly upon guess-work, need hardly be insisted upon" (Weldon, 1895a, p. 379).

This paper was read before the Royal Society in late November of 1894, and in the time between when the paper was read and its eventual publication in February of 1895, Pearson and Bateson both launched broad-scale attacks on Weldon's data, from different directions. Weldon had the opportunity to print a sort of advance rejoinder to their worries, which appeared immediately following the original manuscript. The first section is a response to Bateson, arguing that, with Darwin and Wallace, Weldon's (and Pearson's) gradualist view of evolution is supported by the statistical data that he provides, which demonstrate that even apparently small variations in physiological characters can in fact have significant impacts on trait distributions over time. Weldon then turns to a restatement of the manifesto from the conclusion of the 1893 paper. He explicitly writes that the statistical method means that "all ideas of 'functional adaptation' become unnecessary" to describe the process of evolution (Weldon, 1895b, p. 381), and that similarly, "a theory of the mechanism of heredity is not necessary in order to measure the abnormality of offspring associated with a given parental abnormality" (Weldon, 1895b, p. 381). He closes with a brief response to Pearson, who had urged that the curves did not demonstrate nearly as much as Weldon had argued, by walking back the strength of his conclusions a bit, noting the extent to which the detection of a selective destruction in the crabs depends upon the empirical accuracy of the law of growth which he has ascertained (Pearson was, in turn, extremely doubtful of that law's accuracy; Pearson, 1896).

Let's take stock, then, of the state of the theory around the middle of 1895. Weldon is offering us an extremely spare, statistical picture of evolution. His definition of natural selection is simply a statistical trend in populations of evolving organisms. Accordingly, any depiction of it in the wild, or any experiments designed to detect it, should simply be aimed at the accurate estimation of the relevant statistical constants, along with what other little work is required in order to demonstrate that no alternative, non-selective explanations are applicable in the case at issue. Not only is an account of the physiology of the relevant characters unnecessary, but neither is any real understanding of the material process of heredity required in order to demonstrate the existence of evolutionary change. We need not even clearly understand to what adaptive use the evolutionary changes at work might be directed.

If this sounds a bit radical, and a bit non-Darwinian, it assuredly is. Weldon is soon attacked for this picture, not least because the small "rejoinder" paper, with its particularly provocative restatement of Weldon's 1893 manifesto, was reprinted in *Nature* later in 1895 (Weldon, 1895c). A representative example of the vitriol at work can be found in a letter from the influential British biologist E. Ray Lankester (for more detail on this controversy, see Pence, 2011, forthcoming). Lankester writes, with palpably increasing levels of exasperation, that.

> *Prof. Weldon…has deliberately departed from the simple statement which his observations warranted, viz. that such-and-such a proportion of frontal measurement accompanies survival, and has unwarrantably (that is to say unreasonably) proceeded to speak of the "effect" of this frontal proportion, to declare it to be a* cause *of survival, to estimate the "advantage" and "disadvantage" of this same proportion, and finally to maintain that its "importance" may be estimated without troubling ourselves to inquire how it operates,* or whether indeed it is operative at all.

> *Such methods of attempting to penetrate the obscurity which veils the interactions of the immensely complex bundle of phenomena which we call a crab and its environment, appear to me not merely inadequate, but in so far as they involve perversion of the meaning of accepted terms and a deliberate rejection of the method of inquiry by hypothesis and verification, injurious to the progress of knowledge.*
> **(Lankester, 1896, p. 246, original emphasis)**

Suffice it to say that Weldon had not expected to be declared an enemy of the progress of knowledge, not least because he had, in the period between the publication of the paper in 1895 and this debate in mid-1896, set about a vast program of experimentation on crabs at the Plymouth laboratory, which in part would aim at revising our understanding of the law of growth (as noted above), but would also enable us to see what the physiological function of this frontal breadth in crabs might actually be. Weldon takes himself, in other words, to be engaged in precisely the process of "inquiry by hypothesis and verification" that Lankester is rebuking him for having ignored.

Furthermore, in the intervening time, a significant piece of unpublished data has arrived. Here is Weldon's first discussion of the results with Galton in January 1896, 6 months before the argument in *Nature*. As opposed to his data on female crabs (which had been the subjects in 1893 and 1895),

> *More interesting is an observation of [Weldon's colleague] Herbert Thompson's on* male *crabs from Plymouth. [...] In 1892–3 we had a quantity of crabs from Plymouth.—He took the males, I the females.—This year, he wanted a fresh stock, to complete his series: and at my suggestion kept the 1892–3 set separate from the new set of measures. He finds that in crabs of the same carapace length the frontal ratio is always from 1.5 to 2 units less in the 1895 stock than in the others: so that there is a possibility that the ♂ crabs are slowly diminishing their frontal breadth,—i.e., that their mean is being pushed in the direction indicated by the above hypothesis of selection.*

(Weldon, 1896a, fols. 4r–5r)

Thompson has gone back and compared the original data on male crabs from 1892 and 1893 with a brand new set of data that were collected over the course of 1895; in only 2 to 3 years, there is already a detectable reduction in the male crabs' frontal breadth. In this first note, Weldon is cautious about whether the change will hold up to further statistical analysis, but by the following August he can already confidently claim to Galton that he has found that "the mean frontal breadth of the crabs on the beach below here goes on getting smaller," providing "evidence for a selective destruction of frontal breadths [which] is clear and unequivocal" (Weldon, 1897).

While he has been furthering his statistical research, the physiological strand of Weldon's investigation has been proceeding in parallel. Why, he wonders, might natural selection be in the process of decreasing the frontal breadths of crabs in Plymouth Sound? Weldon writes to Pearson in mid-1896 that he's been working with a colleague on the manner in which a crab drives water across its gills by "wagging parts of its legs" (Fig. 3.5), which is central to his new hypothesis for the selective cause of frontal breadth decrease. "Plymouth Sound is everywhere, and especially near here, becoming yearly muddier, so that the number of dredgers necessary becomes yearly greater, the fauna yearly more scanty," he writes. Such an increase in silt—being man-made, and hence liable to change very rapidly—could, at least potentially, be responsible for the kind of large-scale statistical change that Weldon's crabs show. Further, if frontal breadth is related to some kind of silt-filtering mechanism, involving the pumping of water across or through the area of the crab which frontal breadth describes, then we even have a relatively clear causal story to tell

FIG. 3.5

Weldon's diagram of the role of the frontal breadth of *Carcinus mœnas* in filtering, from a letter to Pearson, August 25, 1896. The arrows indicate places where water can flow.

Credit: Image taken from Weldon's letters, from the Pearson Papers, University College London, call number PEARSON 11/1/22/40.2.1, f. 1v.

which makes frontal breadth a measure of the kinds of characters that could readily be materially affected by the steady silting-in of the sound. "Get on your bicycle," he commands Pearson, "and cultivate a spirit of humble respect for crabs!" (Weldon, 1896b). For Weldon's part, he would only present the full version of this data once, and it is to that address that we now turn.

A settled research program

I want to conclude this chapter by presenting something like the settled "state of play" at which Pearson and Weldon had arrived over the course of the late 1890s. Conveniently for the purposes of our analysis here, both men had the opportunity to publish full presentations of their current research in the span of about a year: Weldon presents an address as the President of Section D (the zoology section) of the British Association on September 8, 1898, and Pearson sends off the second edition of his *Grammar of Science*, with a dramatically expanded section on evolutionary theory covering his new work with Weldon, in December of 1899 (it is published the following month). These two works—despite the fact that both authors complained in correspondence that they wished for more time and space to make them more complete—serve as an excellent lens through which to see what the biometricians thought of their current work and future prospects just prior to the event that would so dramatically reshape all work on evolutionary theory: the rediscovery and reinterpretation of the work of Mendel.

Weldon begins his presidential speech by considering the definition of the theory of natural selection. "In the form in which Darwin stated it," he says, "the theory asserts that the smallest observable variation may affect an animal's chance of survival, and it further asserts that the magnitude of such variations, and the frequency with which they occur, is governed by the law of chance" (Weldon, 1898, p. 887). This much we already know to be false. Darwin had little concept of "the law of chance," and to the extent that he did, he thought of it only as something like the law of large numbers, a principle that guarantees that given enough time and a large enough sample size, even relatively improbable events will eventually occur. While pursuing this thread would take us too far afield here, it is interesting to note that as early as 1898, Weldon is already attempting to use Darwin's *imprimatur* as a weapon in debates over the nature of evolution. The fundamental point, however, that Darwin believed that mutations are not biased in favor of the fitness of organisms, remains true, regardless of whether or not Darwin had any idea what form such a distribution might take.

The frame, then, for Weldon's address is what he calls the "three principal objections" thus far raised to natural selection. Let me consider first the final two—as bearing upon the research in crabs—and close with the first. These are, he writes, "that minute structural variations cannot in fact be supposed to affect the death-rate so much as the theory requires that they should," and that "the process of evolution by Natural Selection is so slow that the time required for its operation is longer than the extreme limit of time given by estimates of the age of the earth" (1898, p. 887). Taken together, they constitute an argument that natural selection could not be responsible for the origin and diversity of life on earth, because small mutations simply cannot alter populations rapidly enough to matter. We require something else—whether large-scale, discontinuous mutations as Bateson thought, or divine creation—to explain the distribution and nature of species.

Both of these objections, he argues, are dealt with by his new data on the Plymouth Sound crabs. As he presents Thompson's data for 1893 and 1895, followed by his own preliminary data for 1898, we see that the mean frontal breadth has decreased quite dramatically. To take just one representative row from his table, in crabs of overall length 13.5 mm, frontal breadths have gone from 10.35 mm in 1893, to 10.31 mm in 1895, to 10.25 mm in 1898—and such decreases have occurred at every single size of crab measured, from 11 to 15 mm. While such differences may appear small, a 1% decrease in any character in a matter of only 5 years is quite remarkable when we consider the time-scales relevant for evolutionary change.

After laying out the transformation of the Sound itself—"the quantity of fine mud on the shores and on the bottom of the Sound is greater than it used to be, and is constantly increasing" (1898, p. 898)—he provides us with a summary of his physiological work. The possibility that the silt was serving as the causal agent, he writes, "induced me to try the experiment of keeping crabs in water containing fine mud in suspension, in order to see whether a selective destruction occurred under these circumstances or not" (1898, p. 899). The negative influence of silt on frontal breadth is confirmed by every experiment he is able to run. In a wide array of crabs exposed to silt, those with narrower frontal breadths are more likely to survive than those with broader "foreheads" (the mean frontal breadth of survivors is reduced). The same effect occurs both with clay and with very fine mud. And, though the experiment is difficult to perform and the conclusions are thus speculative, populations of crabs reared in an artificial environment, entirely without silt, appear to have significantly broader average frontal breadths than the wild population. The process, he writes, "seems to be largely associated with the way in which crabs filter the water entering their gill-chambers" (1898, p. 901)—the crabs killed by silt having their gills covered in a fine layer of mud.

We are thus entitled to the following two conclusions about the Plymouth Sound crabs, Weldon writes:

> *The first is that their mean frontal breadth is diminishing year by year at a measurable rate, which is more rapid in males than in females; the second is that this diminution in the frontal breadth occurs in the presence of a material, namely, fine mud, which is increasing in amount, and which can be shown experimentally to destroy broad fronted crabs at a greater rate than narrower frontal margins.*
>
> **(1898, p. 900)**

On this basis, he continues, "I see no escape from the conclusion that we have here a case of Natural Selection acting with great rapidity because of the rapidity with which the conditions of life are changing" (1898, p. 900). Weldon is thus confident—as have been many commentators to follow him in the century hence—that this is the first demonstrated case of the operation of natural selection in the wild. The first two objections to selection are overthrown, and Darwin's theory is no longer hypothetical, or confined to the realm of the unobservable.

But this leaves the first objection: that, because "the species of animals which we know fall into orderly series," and orderly series would be unlikely to be produced by chance, "the variations on which the process of Natural Selection had to act must have been produced by something which was not chance" (1898, p. 887). One imagines that Weldon has several kinds of potential opponents in mind here. Some might be those like St. George Jackson Mivart, who in his *On the Genesis of Species* argued for the divinely directed character of organic variation. More important, however, would be objectors like J. T. Cunningham, one of Weldon's regular sparring partners in the correspondence pages of *Nature* (Cunningham, 1896, 1895), a prominent neo-Lamarckian, the translator of some of the important works of orthogenesis into English (Eimer, 1890), and Weldon's colleague at the Plymouth laboratory. The debate here concerns two factors, often interrelated, if not confused (Bowler, 1992; Gayon, 1992)—first, are any characters that are acquired by an organism during its lifetime transmitted on to its offspring? Darwin believed that they were, at least in some cases, though as mentioned in the last chapter, the burgeoning dispute over the work of Weismann has since begun to place this in doubt. Second, are there laws that govern the distribution of variations? That is, does the fact that variation tends to cluster along certain "axes"—with some parts more liable to vary more often, and in particular directions, than others—mean that we must posit particular "laws" of growth to explain this directional change? Weldon, for his part, is no great believer in the inheritance of acquired characters (at least in the neo-Lamarckian sense), arguing rather on Galtonian grounds that every character is in part present in the inherited material which an organism receives from its parents and more remote ancestors, and in part enabled by environmental and developmental interaction (about which more in the next chapter). And he clearly has no use whatsoever for orthogenetic laws of development, as they contradict the evident applicability of the statistical study of variation.

Before turning to Weldon's argument, let's remember the stakes. Our imagined objector is claiming that variation is not, in fact, distributed according to the law of chance. This is thus no trifling complaint—it would upend the entire possibility of the statistical study of evolution in the way that Weldon and Pearson have conceived it. Weldon responds to the worry by, for the first time, working to define his terms. "The meaning of the word Chance," he writes, "is so thoroughly misunderstood by a number of writers on evolution that I make no apology for asking you to consider what it does mean" (1898, p. 887). For Weldon at this stage, chance is a description of a very particular variety of ignorance. In some types of events in nature, he argues, "we know so little about a group of events that we cannot predict the result of a single observation, although we can predict the result of a long series of observations." In such cases, "we say that these events occur by chance" (1898, p. 888). The tossing of dice is like this—as he demonstrates using some 16,000 rolls of dice performed by his wife. Such tosses are distributed according to perfectly predictable laws (though not quite normally distributed, as the irregularity of the dice and the small bits of material removed for the pips render them slightly asymmetrical), and variations work in exactly the same way. Notably, this is not an a priori claim: rather, it is a fact to be demonstrated by empirical and mathematical study. And the mathematical tool best cut out for the job isn't Galton's normal distributions, but rather the method for handling asymmetric curves pioneered by Pearson in his 1895 memoir which appeared shortly following Weldon's first paper on crabs. The ability to construct asymmetric curves (along with Pearson's work on the hypergeometric distribution, which appears in the same paper) "made it possible to apply the theory of chance to an enormous mass of material, which no one had previously been able to reduce to an orderly and intelligible form" (1898, p. 892).

Why is the ability to consider asymmetric curves so vital for Weldon's work? Because here he finally finds a solution to the problem of Galtonian evolutionary stasis. For should a character—take longer frontal breadth in crabs—become advantageous to an organism, then the organisms with the longest frontal breadth will reproduce more than their short-breadth conspecifics, steadily producing a longer "tail" on the character distribution, in a manner that we actually see in Weldon's crab data. In a time-frame as short as a single generation, we could see significant movement in the mean value for the trait within the population. "This view of the possible effect of selection," he writes, "seems to have escaped the notice of those who consider that favourable variations are of necessity rare, and likely to be swamped by intercrossing when they do occur" (1898, p. 895). Here is the way to make continuous variation compatible with natural selection, without the threat of favorable variations being drowned out by the less-fit members of the population.

Weldon is perceptive enough to realize that this will not always occur, however. He proceeds to present two cases in which (to use modern phrasing) available variation limits the response to selection, one drawn from pigs in which variation in one direction seems to be much more difficult than that in another, and one drawn from poppies in which the reduction of a character below a threshold value seems to be impossible. But, and here is Weldon's most trenchant point, the knowledge of data concerning distribution of variations is what gives us the ability to make such predictions, which had not before been possible. It is therefore, he argues, "the duty of students of animal evolution to use the new and powerful engine which Professor Pearson has provided, and to accumulate this kind of knowledge in a large number of cases" (1898, p. 896). Not only is the objection to natural selection thus resolved, but it is resolved in a way that provides novel insight previously unavailable to the working biologist.

Now, this is not to say that there will not be criticism. "I know that there are people," he continues, "who regard the mode of treatment which I have tried to describe as merely a way of saying, with a pompous parade of arithmetic, something one knew before" (1898, p. 896). Such a person should, however, be convinced by the sorts of theoretical advances which have already resulted from work like his own and that of Galton, he claims.

Here, then, is Weldon's picture in mid-1898. Not only has natural selection now been demonstrated in the wild, and at such a rate as to make it possible for it to be a significant factor in evolutionary change, but it has set up an entire research program for biometry. The collection of data on variation in the wild (and Weldon is, at this point, pursuing such data in sources as diverse as microorganisms, snails, sparrows' eggs, poppies, and more) can give us an idea of the raw material available for a response to selection, and careful monitoring of population distributions over time can lead us to hypotheses about the physical causes that could be responsible for such selection. Biometry has a mission. The closing words of Weldon's address are worth quoting at some length:

> The whole difficulty of the theory of Natural Selection is a quantitative difficulty. It is a difficulty of believing that in any given case a small deviation from the mean character will be sufficiently useful or sufficiently harmful to matter. [...] We ought to know numerically, in a large number of cases, how much variation is occurring now in animals; we ought to know numerically how much effect that variation has upon the death-rate; and we ought to know numerically how much of such variation is inherited from generation to generation. The labours of Mr. Galton and of Professor Pearson have given us the means of obtaining this knowledge, and I would urge upon you the necessity of obtaining it. For numerical knowledge of this kind is the only ultimate test of the theory of Natural Selection, or of any other theory of any natural process whatever.
>
> **(Weldon, 1898, p. 902)**

Pearson's view of this mission—in interesting ways both a continuation of and contrast with Weldon's—is presented in the second edition of his *Grammar of Science*. While I have been relatively quiet about Pearson's contributions thus far—predominantly because they are so technical that presenting them in any significant detail would drastically alter my project—Pearson's role in the development of biometry should not be understated. The two men often attacked one another, Pearson going after Weldon for mathematical naivety and Weldon objecting to Pearson's lack of biological sensitivity, but their collaboration ran deep. While he retains the original, non-evolutionary chapter on the relation between life and physics from the first edition of the *Grammar*,

Pearson adds two new chapters on evolution, crediting Galton, Weldon, and G. Udny Yule, Pearson's "demonstrator," collaborator, and laboratory technician at University College from 1893 until 1899 (and who will return to our narrative in Chapter 5). These constitute some 125 pages laying out Pearson's understanding of evolution.

While Pearson thus doesn't repudiate any of the work on life that he had written a decade prior, he does start his first chapter on evolution by immediately recognizing its shortcomings. "In the last chapter," he writes, "we freely used the words 'evolution' and 'selection' as if they had current common values. Now this is very far from being the case…" Indeed, he argues, they could not have had clear definitions when he wrote the book's first edition in 1892, for "it is only within the last few years, however, with the growth of a quantitative theory of evolution, that precise definition of fundamental biological concepts has become possible" (Pearson, 1900, p. 372). Non-quantified, or non-quantifiable, approaches to the life sciences, he argues—and here he includes as an explicit example Weismann's work on the germ plasm—"is imagination solving the universe, propounding a formula before the facts which the formula is to describe have been collected and classified" (1900, p. 373). Of course, it makes perfect sense that Pearson would approach evolution in such a way. As noted above, his positivism prized above all the depiction (and, by extension, the simplification) of the world in terms of straightforward mathematical laws, a theme to which we will return shortly in discussing Pearson's emphasis on a reworking of Galton's Law of Ancestral Heredity. From the very beginning of his introduction of evolution, he sets for himself the following challenge: "Under what formula shall we economise thought when we attempt to describe scientifically this vast field?" (1900, p. 374). We have here, then, Pearson's basic positivist methodology, specialized for the particular case of evolutionary theory. Just as in the physical sciences, "the object of the naturalist is…to replace the longer and more complex descriptions by more comprehensive and simpler descriptions—to discover…facts which can be described by a few formulae, nay, if possible, by one brief formula" (Pearson, 1900, pp. 500–501).

And what is the methodology by which we can effect this simplification in the case of evolution? Unsurprisingly, it is correlation between distributions of traits. In fact, the vast majority of his chapter on evolution is, using a combination of real and contrived biological examples, an introduction to the mathematics of statistics and regression. As he puts the matter, "when we say that evolution is taking place, we mean that progressive changes are going on in one or all of the numerical values which fix the mean, variability, and correlation of the system of organs and characters" (1900, p. 403). He thus introduces the concepts of distributions, with their properties of mean, mode (which he has recently defined), and standard deviation (the term, though not the concept, another Pearson coinage), moving on to correlation between multiple distributions, first of the same character measured in common units, then between different characters measured in units of deviation from the mean. When it comes to the measurement of selection itself, he essentially recapitulates Weldon's method in his crabs, though with artificial data. The mere existence of selection is, at this point, beyond doubt, he writes. "To assert that natural selection does not exist is to assert that the whole death-rate is non-selective; or that it is not a function of the constitution, the characters and organs of the individual. Looked at from this standpoint, every medical practitioner, every careful observer of nature, has seen selection at work" (1900, p. 496). Measuring it, however, is another matter. Since we can't easily isolate or track individual members of a population over any reasonable length of time in the wild (one is reminded of Weldon's objection to Galton's proposed experiment on shrimp mentioned at the beginning of the chapter), we can't get perfectly precise measures of selection. What we can do by way of estimating it, however, "is to measure, say a sample 1000 adult members of a population at one time and a second sample 1000 at a later time" (1900, p. 408), exactly as Weldon has done in Plymouth Sound. He closes with a note of optimism about future prospects: "In a few years we may hope no longer to hear natural selection spoken of as hypothetical, but rather to listen to a statement of its quantitative measure for various organisms under divers[e] environments" (1900, p. 500).

The great unsolved questions in evolution, then, concern the form and strength of this selection. Can it actually produce reproductive isolation, and hence be responsible for the creation of species? And can it occur rapidly enough that it could be the mechanism that was in fact responsible for the species that we see on earth (Pearson, 1900, p. 412)? This requires a detailed analysis of the possible types of selection and the various mechanisms that might cause reproductive isolation—notably, for Pearson, not in terms of biological or physiological details, but rather in terms of how we might statistically detect the signal of processes like preferential mating. Pearson's discussions of the mechanics of heredity are very nearly plagiarized from Galton's work in the late 1870s (1900, p. 456), though of course with an extreme level of mathematical precision present throughout. One further particularly striking difference between Weldon and Pearson becomes apparent in these two chapters: Pearson's examples are almost entirely drawn from humans, consistent with his emphasis on eugenics. His worries about fertility, preferential mating, and so on, are immediately connected to worries of class and race in England.

One final Pearsonian innovation is worthy of note here, given the importance which both Pearson himself and, later, his son Egon would ascribe to it. Following on a paper that Pearson had prepared as "a New Year's Greeting to Francis Galton" (Pearson, 1898), he presents his efforts to rework and update Galton's original Law of Ancestral Heredity, which was mentioned briefly in the last chapter. Galton had—on the basis of some perhaps dubious empirical and theoretical assumptions (Bulmer, 1998)—derived a formula for the fractions of heritable characters in each organism which could be traced to parents, to grand-parents, and so forth for further generations back into more remote ancestry. Pearson sees here the possibility of precisely the kind of powerful, descriptive law that his philosophy of science so deeply valued. "If Darwinism be the true view of evolution, *i.e.* if we are to describe evolution by natural selection combined with heredity, then the law which gives us definitely and concisely the type of the offspring in terms of the ancestral peculiarities is at once the foundation-stone of biology and the basis upon which heredity becomes an exact branch of science" (Pearson, 1900, p. 479). Pearson offers a more general form of Galton's result (which takes the form of Galton's equation after simplification and the choice of a constant). He also considers how we might express such a law not only for the case of continuously varying characters, which are more amenable to being seen as built from fractions of values found in ancestors, but also for discretely varying characters (like eye-color), which are more challenging to treat statistically. Such a law, Pearson argues, would offer many further consequences worthy of test—for example, values for coefficients of relationship not only between organisms and their ancestors, but organisms and their extant relatives could be derived (e.g., between an individual person and their siblings or cousins).

In short, Pearson's project for biometry centers around the statistical classification of the various types of selection that may be at work, in order to determine the extent to which natural selection might possibly play the role which Darwin had described for it. Only with a clear idea of the way in which we might express various types of selection can we hope to properly construct a theory of evolution, which for Pearson therefore means we require knowledge of the statistical forms of the multiple influences of factors like preferential mating, self-fertilization, or non-selective death-rates, and the ways in which resulting characters could be passed down the generations. The broader project thus fits perfectly with Weldon's. Weldon's obsession with data concerning variation in natural populations provides raw material needed both to calibrate Pearson's mathematical derivations to the natural world, and to determine which experimentally tractable populations in the wild would actually serve as promising targets of future study. The goal, then, is to steadily make this enterprise more sophisticated, teasing apart the various sub-species of natural selection and slowly demonstrating that it could, in fact, be the mechanism responsible for the origin of species.

This was the research program of biometry as a new century dawned in 1900. It is an interesting exercise in counterfactual history to consider what biological science might have been like had it ever been put into practice. But, in fact, it was not. Over the course of 1900, as is by now well known, Carl Correns, Erich von Tschermak, and Hugo de Vries would "rediscover" the work of Gregor Mendel on the inheritance of characters in peas—and the science of evolution and heredity would be permanently altered.

References

Bateson, W., 1894. Materials for the Study of Variation, Treated with Especial Regard to Discontinuity in the Origin of Species. Macmillan, London.

Bowler, P.J., 1992. The Eclipse of Darwinism: Anti-Darwinian Evolution Theories in the Decades around 1900. Johns Hopkins University Press, Baltimore, MD.

Bulmer, M., 1998. Galton's law of ancestral heredity. Heredity 81, 579. https://doi.org/10.1046/j.1365-2540.1998.00418.x.

Clifford, W.K., Pearson, K., 1885. The Common Sense of the Exact Sciences. D. Appleton and Company, New York.

Cock, A.G., Forsdyke, D.R., 2008. Treasure Your Exceptions: The Science and Life of William Bateson. Springer, New York.

Cunningham, J.T., 1895. The statistical investigation of evolution (letter of Mar. 28, 1895). Nature 51, 510. https://doi.org/10.1038/051510a0.

Cunningham, J.T., 1896. The utility of specific characters (Letter of July 30, 1896). Nature 54, 295. https://doi.org/10.1038/054295a0.

Eimer, T., 1890. Organic Evolution as the Result of the Inheritance of Acquired Characters According to the Laws of Organic Growth. Macmillan, London.

Gayon, J., 1992. Darwin et l'après-Darwin: une histoire de l'hypothèse de sélection naturelle. Éditions Kimé, Paris.

Gould, S.J., 1971. D'Arcy Thompson and the science of form. New Lit. Hist. 2, 229–258. https://doi.org/10.2307/468601.

Lankester, E.R., 1896. Are specific characters useful? (letter of Jul. 16, 1896). Nature 54, 245–246. https://doi.org/10.1038/054245c0.

Mach, E., 1919. The Science of Mechanics: A Critical and Historical Account of its Development, fourth ed. Open Court, Chicago and London.

Pearson, K., 1889. On the Laws of Inheritance According to Galton.

Pearson, K., 1893. Contributions to the mathematical theory of evolution. [Abstract]. Proc. R. Soc. Lond. 54, 329–333.

Pearson, K., 1894. Contributions to the mathematical theory of evolution. Philos. Trans. R. Soc. Lond. A 185, 71–110. https://doi.org/10.1098/rsta.1894.0003.

Pearson, K., 1896. The utility of specific characters (letter of Sep. 17, 1896). Nature 54, 460–461. https://doi.org/10.1038/054460c0.

Pearson, K., 1898. Mathematical contributions to the theory of evolution. On the law of ancestral heredity. Proc. R. Soc. Lond. 62, 386–412. https://doi.org/10.1098/rspl.1897.0128.

Pearson, K., 1900. The Grammar of Science, second ed. Adam and Charles Black, London.

Pearson, K., 1906. Walter Frank Raphael Weldon. 1860–1906. Biometrika 5, 1–52. https://doi.org/10.1093/biomet/5.1-2.1.

Pearson, E.S., 1936. Karl Pearson: an appreciation of some aspects of his life and work. Part I: 1857–1906. Biometrika 28, 193–257. https://doi.org/10.2307/2333951.

Pence, C.H., 2011. "Describing our whole experience": the statistical philosophies of W. F. R. Weldon and Karl Pearson. Stud. Hist. Philos. Biol. Biomed. Sci. 42, 475–485. https://doi.org/10.1016/j.shpsc.2011.07.011.

Pence, C.H., forthcoming. How not to fight about theory: the debate between biometry and Mendelism in Nature, 1890–1915, in: De Block, A., Ramsey, G. (Eds.), The Evolution of Science. University of Pittsburgh Press, Pittsburgh, PA.

Porter, T.M., 2004. Karl Pearson: The Scientific Life in a Statistical Age. Princeton University Press, Princeton, NJ.

Provine, W.B., 1971. The Origins of Theoretical Population Genetics. Princeton University Press, Princeton, NJ.

Radick, G., 2005. Other histories, other biologies. R. Inst. Philos. Suppl. 56, 21–47. https://doi.org/10.1017/S135824610505602X.

Sloan, P.R., 2000. Mach's phenomenalism and the British reception of Mendelism. C. R. Acad. Sci. III 323, 1069–1079. https://doi.org/10.1016/S0764-4469(00)01255-5.

Thiele, J., 1969. Karl Pearson, Ernst Mach, John B. Stallo: Briefe aus den Jahren 1897 bis 1904. Isis 60, 535–542. https://doi.org/10.1086/350540.

Thiselton-Dyer, W.T., 1895. Variation and specific stability (letter of Mar. 14, 1895). Nature 51, 459–461. https://doi.org/10.1038/051459c0.

Weldon, W.F.R., 1890a. The variations occurring in certain decapod Crustacea.—I. *Crangon vulgaris*. Proc. R. Soc. Lond. 47, 445–453. https://doi.org/10.1098/rspl.1889.0105.

Weldon, W.F.R., 1890b. Letter from WFRW to FG. 1890-01-07.

Weldon, W.F.R., 1890c. Letter from WFRW to FG. 1890-01-23.

Weldon, W.F.R., 1890d. Letter from WFRW to FG. 1890-02-16.

Weldon, W.F.R., 1890e. Letter from WFRW to FG. 1890-02-19.

Weldon, W.F.R., 1892a. Letter from WFRW to KP. 1892-11-27.

Weldon, W.F.R., 1892b. Letter from WFRW to FG. 1892-11-27.

Weldon, W.F.R., 1892c. Certain correlated variations in *Crangon vulgaris*. Proc. R. Soc. Lond. 51, 2–21.

Weldon, W.F.R., 1893. On certain correlated variations in *Carcinus mœnas*. Proc. R. Soc. Lond. 54, 318–329. https://doi.org/10.1098/rspl.1893.0078.

Weldon, W.F.R., 1893. Letter from WFRW to FG. 1893-07-14.

Weldon, W.F.R., 1894. The study of animal variation [review of Bateson, W., *Materials for the Study of Variation*]. Nature 50, 25–26. https://doi.org/10.1038/050025a0.

Weldon, W.F.R., 1895a. An attempt to measure the death-rate due to the selective destruction of *Carcinus mœnas* with respect to a particular dimension. Proc. R. Soc. Lond. 57, 360–379. https://doi.org/10.1098/rspl.1894.0165.

Weldon, W.F.R., 1895b. Remarks on variation in animals and plants. Proc. R. Soc. Lond. 57, 379–382.

Weldon, W.F.R., 1895c. Variation in animals and plants. Nature 51, 449–450. https://doi.org/10.1038/051449a0.

Weldon, W.F.R., 1896a. Letter from WFRW to FG. 1896-01-05.

Weldon, W.F.R., 1896b. Letter from WFRW to KP. 1896-08-25.

Weldon, W.F.R., 1897. Letter from WFRW to FG. 1897-08-04.

Weldon, W.F.R., 1898. Address of the President of Section D (zoology). Rep. Br. Assoc. Adv. Sci. 68, 887–902.

Here is the true gospel: Biometry after Mendelism

It is easy to say Mendelism does not happen. But what the deuce does happen is harder every day!
Weldon, letter to Pearson, 3 March 1903

On the 16th of October 1900, Weldon wrote to Pearson, bemoaning political intrigue at University College. He then abruptly changed the subject:

About pleasanter things, I have heard of and read a paper by one Mendel on the results of crossing peas, which I think you would like to read. It is in the Abhandlungen des Naturforschenden Vereines in Brünn *for 1865—I have the R. S. [Royal Society] copy here, but I will send it to you if you want it.*

The point seems to me to be that the results indicate an exclusive inheritance with a very high paternal correlation.

(Weldon, 1900a, fols. 1r–1v)

Weldon goes on to describe the bulk of Mendel's results. The discussion isn't worth recapitulating to anyone who has a faint memory of their secondary-school biology course, but one thing that makes it particularly notable here is the ease with which it fits straightforwardly into Weldon's biometrical picture of inheritance. Weldon formalizes the ratios of traits that would be expected after multiple generations, and closes by noting that "it seems a very good starting point for further work" (Weldon, 1900a, fol. 3v). In short: business as usual. Mendel's paper presents a particularly interesting pattern of correlation in a number of characters of interest in peas, and Weldon plans to follow up on it in the manner we would have expected given the plan for future studies laid out in the last chapter. Further letters over the next few months tease out the sense in which Mendel refers to "generations" (essential for comparing the various efforts of Mendel's "rediscoverers" to reproduce his experiments), along with the notion of "hybrid" at work in Mendel's theoretical picture.

By early December, however, it is clear that Mendel will be rapidly enlisted as a weapon by critics of Weldon and Pearson's statistical methodology. As Weldon pens in the postscript to a letter to Pearson, "it is absurd to say that Mendel upsets the result of other observations, or destroys the value of a way of treating them.—That is all part of the assumption that there is only one man in England (or elsewhere)

The Rise of Chance in Evolutionary Theory. https://doi.org/10.1016/B978-0-323-91291-4.00004-2

[Bateson, that is] who has any insight into the workings of Nature" (Weldon, 1900b, fols. 4r–4v). To get an idea of how this early conflict developed, consider a skirmish that would prove unusually pivotal: the publication of Pearson's paper "Mathematical contributions to the theory of evolution. IX. On the principle of homotyposis and its relation to heredity, to the variability of the individual, and to that of the race," which would appear in the *Philosophical Transactions* in early 1901. Prior even to its acceptance, Bateson had received a copy in manuscript, and managed to present a public critique of it, read at the Royal Society. Pearson, of course, could not modify the article, despite the fact that he believed Bateson's criticisms to reflect easily avoided misreadings—to do so would be to make Bateson's assessment, already now read into the record, seem foolish. And yet the paper would be printed practically "pre-refuted," packaged only some few pages away from a critical response. Pearson is furious, but his hands are tied.

Bateson's critique focuses—in line with his views as sketched in the last chapter—on what he believes to be the discontinuous nature of important, evolutionarily relevant variations. Because statistical analysis combines discontinuous and continuous variation, it can never, Bateson thinks, properly describe evolutionary phenomena. "[M]uch of the statistical work produced by Professor Pearson and his followers has, I believe, gone wide of its mark," he argues, "if that aim is the elucidation of Evolution. More fitly might this work be described as 'Mathematical Contributions to a Theory of Normality'" (Bateson, 1901, p. 203). He clarifies later that "any attempt to understand variation in its relation to Evolution" (Bateson, 1901, p. 204) must take account of the distinction between these two types of variability. Mendelism, then—with its stark differentiation between green and yellow, smooth and wrinkled peas—appears to Bateson to be precisely the sort of theory that could ground the appearance of variations that might actually lead to the production of new species. By March 1902, Bateson would deploy Mendel as a way to directly and personally attack Weldon's "haste to annul this first positive achievement of the precise method," in favor of Mendel, who, "alone, and unheeded, [had] broken off from the rest—in the moment that Darwin was at work—and cut a way through" all open problems in the entirety of inheritance and evolution (Bateson, 1902, pp. xi, v). (Bateson's insistence here that Weldon is "disposed to criticism rather than to cordiality" [1902, p. 116] is a fine instance of projection.) But this takes us too far forward too quickly and mires us in the details of Bateson and Weldon's personal feud. Let's double back to consider some further context and then return to Weldon's theorizing after the arrival of Mendelism.

New ports in new storms

Even in the absence of the arrival of Mendel's paper and the furor which it generated, this was a time of great institutional upheaval for Pearson and Weldon, which in part explains their novel theoretical work to which we will soon turn. One such development had already taken place.

Weldon had departed University College for a professorship at Oxford in February of 1899. While the prestige of an Oxford post was too important a career move for Weldon to pass up, he always lamented the fact that he had a much harder time recruiting quality students in Oxford than he did in London.[a] For his part, Pearson is heartbroken; he writes to Weldon—in a letter marked "don't answer this"—that "you have changed the whole drift of my work and left a far deeper impression on my life than I on yours" (Pearson, 1899, fol. 2r). The lunchtime discussions that had been so productive between the two came to an abrupt end, and the pair now focused their efforts on holidays spent collecting data in the countryside.

A second major professional development for Weldon and Pearson is a further consequence of Bateson's machinations at the Royal Society. The Evolution Committee—founded, as we saw in the last chapter, as a refuge for biometrical work—had been opened to Bateson and his allies as a conciliatory gesture. The Batesonians, however, even prior to the rediscovery of Mendel, had begun to take an increasing number of the committee's positions and to redirect an ever larger share of its funds toward their projects and away from biometrical work. The steady drift (or, so it appeared to the biometricians) of the Royal Society's journals toward Batesonian and then Mendelian publications would cement Weldon and Pearson's growing belief that the Royal Society as a whole was becoming an institution hostile to their program. After Weldon's death, Pearson would write to Galton, at this point so frustrated with the society that he is considering renouncing his fellowship, saying it was something that "Weldon & I more than once talked about" (Pearson, 1906a, fol. 5).

We thus return to the tale with which I opened the book. A bit less than a month after writing to Pearson about his first reading of Mendel's paper, Weldon wonders:

> *Do you think it would be too hopelessly expensive to start a journal of some kind? I fancy that you could about fill a sufficiently* selten erscheinende Zeitschrift *[seldom-published journal] yourself. [...] I think an* Archiv für K.P.ismus *would become a thing to be reckoned with. It ought to contain original papers on very strictly limited lines, and abstracts of biological papers bearing on the subjects treated.*

(Weldon, 1900c, fols. 1v–2r)

By early 1901, plans are in full swing, and the bulk of the correspondence between Pearson and Weldon is now devoted to matters of the newly christened *Biometrika*. After much drama involving publishers, printing rates, promises of subscriptions at home and abroad, and the like, the first issue of the journal appears in October. With Weldon, Pearson, and their colleague Charles B. Davenport (an English-trained American biometrician) as editors, working officially "in consultation with Francis Galton," the issue begins with a photograph of the sculpture of Darwin in the

---

[a] Indeed, it is likely that some of the sociological dimension of the long-term debate that would unfold between the biometricians and the Mendelians is due to Bateson's greater facility in obtaining motivated students from his Cambridge position.

University Museum at Oxford, fashioned in the shape of a religious icon, followed by the motto *Ignoramus, in hoc signo laboremus*—we are ignorant; in this sign we work.

The first article is an editorial defining the journal's scope. "A few years ago," the editors write, "all those problems which depend for their solution on a study of the differences between individual members of a race or species, were neglected by most biologists" (Weldon et al., 1901, p. 1). Only with statistical method and inference can we hope to make progress—in particular, by giving statistical researchers the opportunity to publish not only their final results and conclusions, but also to present the data upon which they are based (a privilege not often allowed in the Royal Society journals and never permitted in a more space-constrained periodical like *Nature*). They hoped, moreover, to instill a normative standard for good statistical data collection via their editorship, "a uniformity of statistical treatment, terminology, and notation, so that results obtained by different investigators on different types of life may be easily and effectively compared" (Weldon et al., 1901, p. 2).

Thus, then, were the translation of Mendel's paper and the renewed, forceful attack of Bateson's group placed directly into a context of extreme personal and professional upheaval for the biometricians. Weldon's vision for the future of biometry, having been refined over the last years of the 1890s, was soon to change significantly.

From Mendel to inheritance

Before the end of 1900, we already see hints of this new direction in Weldon's research. He writes to Pearson, focused suddenly on the mechanisms of inheritance that could give rise to the kinds of statistical patterns that the biometricians had been exploring, as well as those described by Mendel.

> *I think that there must be an element in each gamete corresponding to every quality transmitted by it; some of these may blend with the corresponding elements of the other, some may exclude corresponding elements of the other, some may make a patch work resulting in a particulate [i.e., variegated or mosaic] inheritance.*
>
> **(Weldon, 1900d, fol. 1r)**

Weldon has thus, in a matter of weeks, settled upon the approach to Mendel's results that would characterize his work on the question for the remainder of his life. He realizes that he has yet to repair one of the most significant holes in Galton's research program: the process by which individual characters are taken from parents and recombined into offspring. Recall that Galton himself had very clearly drawn the distinction between patent and latent characters (which could, in turn, explain the phenomenon of reversion to remote ancestors) and argued that expression of characters within an organism would have, at least in part, to do with the location of those characters and with their relationships (about which more in a moment, via Weldon's emphasis on the same point). But beyond this, he had only made vague political analogies about a form of "selection" for patent characters which would take place when offspring germ cells were produced in parents. Weldon's goal is to bridge this gap and to give us an account by which characters could be traced from the elements present in parental cells to those in offspring cells. Such inter-generational distributions, he hoped, would produce a theory that could both take account of results like those that he had already obtained for his crabs (which were apparently common instances of blending inheritance that presented nothing like the phenomenon of Mendelian dominance) and still render Mendelian results as a special case. The question, then, is how to mathematically describe and parameterize this assortment of elements from parents into offspring (eventually, via the intermediate step of the formation and division of chromosomes).

As an aside, I should note that we have a persistent problem of terminology here. Weldon is drawing on Galton's work on the material basis of heredity, which Galton referred to, after the failure of pangenesis, with his new term "stirp." Weldon, for his part, does not often use that term, and clearly imputes it to Galton when he does. He writes that neither does he "like the gemmule notation, because it ties one up to Darwin's pangenesis, and I do not see that this is necessary" (Weldon, 1900d, fol. 1v). In fact, he would not settle on a particular term for these units of character transmission before his death (switching in his later notebooks, e.g., to "chromomeres," to indicate their status as components of chromosomes); I am thus forced to call them "elements," with full knowledge that this is just one among many ways in which Weldon discussed the relevant category.

This thread of research becomes a steady background in Weldon's notes and correspondence, throughout an ensuing period of extremely intense work collecting data on a variety of species for biometrical study, and, in particular, a very demanding editorial program for his and Pearson's young journal. We next check in on his progress the following year, in May 1901. His thinking about the process of segregation has led him to consider what might happen when mixed forms of the assortment of elements occur over multiple generations of transmission of the same character:

> It seems to me fairly clear that Mendel's results include two sets of things which have not necessarily any connection with each other.

> If you have two races, or species, one with a character A, the other with the corresponding character α, and cross them, the hybrids may all be A, or they may all be α, or they may be intermediate between A and α, or they may be a patchwork of α and A.

> This seems to me a result of mixing two gametes of different constitution so as to form a body: and the character of that mixture need not necessarily, for any a priori reason that I can see, affect the question whether these elements will be mixed in the gametes produced by the hybrid body, or not: so that characters may at least conceivably blend in a hybrid body of the first generation, and segregate in the gametes of that body, in which case Mendel's law of subsequent segregation would hold, but his law of dominance would not. On the other hand, a character, which is dominant when the elements of 2 different gametes are forcibly mixed

by hybridisation, ought plausibly to be dominant also if the same elements were mixed in the single gametes produced by the hybrid body: so that the law of dominance might hold in hybrids which "breed true."

(Weldon, 1901, fols. 1v–2v)

This letter is worth quoting at length, because it shows, perhaps most clearly of any of Weldon's writings, the extent to which he is seeing Mendelism as just one possibility among the myriad ways in which we might combine the elements of parents to produce their offspring. As he describes here, Mendelian inheritance, showing dominance of one character over others in the first hybrid generation, followed by the clear assortment of organisms in the following generation into two categories, with three-quarters dominant and one-quarter recessive, is in fact the result of very particular methods of assortment and inheritance (as well as, in Weldon's terms, the activation of latent vs patent elements), spread out over multiple generations. At the very least, there are three broad types of inheritance that we might imagine—what Weldon calls blending, alternative, and particulate inheritance. The first is the one most thoroughly treated by biometrical study, and can be found, for instance, in his crab work. Particulate inheritance (as a pattern of phenotypic transmission, not to be confused with any hypothesis about the elements themselves) describes the production of variegated offspring, where characters mix, but remain individually distinct. Alternative inheritance, finally, is that described by Mendel, in which an organism takes on one character or the other of a pair, as mutually exclusive alternatives.

Even if we can get clear on what kind of inheritance is expressed in some case—that is, on the relationship between the elements carried in the gametes and the visible character traits expressed by the adult formed from them—this need not be connected to the manner in which the gametes of the next generation are, in turn, formed. The more Weldon tries to construct a general theory, the more choices he realizes must be made, and this in the absence of the kind of data which the biometricians always sought prior to grand theorizing. "The whole thing," he writes, "is very fascinating, but it is like all the other things. One cannot really know anything about it from books, and one cannot find out otherwise in less than several lifetimes" (Weldon, 1901, fol. 2v).

Another important avenue for Weldon consists in elucidating just what Mendel's experiments purport to show in the first place. As has been extensively discussed by Gregory Radick (2011, 2015), Weldon is the first person to engage in a comprehensive reanalysis of Mendel's data. While he does not—contrary to the later work of Fisher—accuse Mendel of fraud, noting only the "wonderfully consistent way in which Mendel's results agree with his theory" (Weldon, 1902a, p. 232), he is spurred by consulting Tschermak's data on varieties of peas different from those that Mendel used (and offering a much less precise instantiation of Mendel's laws) to launch a thorough reconsideration of the phenomena of dominance and the ways in which Mendel has selected his relevant characters.

The results would be published in a long article in *Biometrika*, often now described—rather unfairly, as we will see—as a broadside attack on Mendel and his results. The framing of the paper is entirely in keeping with what we have seen of

Weldon's unfolding discussions with Pearson. While the general work of the biometrical school has given us a solid understanding of blended inheritance, our knowledge concerning alternative and particulate inheritance, he writes, "is however still rudimentary, and there is so much contradiction between the results obtained by different observers, that the evidence available is difficult to appreciate." Consistent with the view of Mendel's results as a complementary advance fitting squarely within the research tradition of the biometricians, then, Weldon describes the purpose of his paper as "to describe some cases of alternative inheritance, which have lately excited attention" (Weldon, 1902a, p. 228).

Weldon brings together not only Tschermak's somewhat non-Mendelian results, but also work by de Vries, Correns, and observations of some dozen or so other breeds of peas that have been cultivated over the years from common stock by producers of seed catalogs—whose archives constitute a good source of data about the stability of these lineages of peas over multiple generations. Notably, he does not cast any doubt on the validity of Mendel's observations. Rather, it is the fact that they are valid which leads to the problem. For "the question at once arises, how far the laws deduced from [Mendel's observations] are of general application. It is almost a matter of common knowledge that they do not hold for all characters, even in Peas, and Mendel does not suggest that they do. At the same time I see no escape from the conclusion that they do not hold universally for the characters of Peas which Mendel so carefully describes" (Weldon, 1902a, p. 235). While they may have held in Mendel's own strain of peas, that is, a broader lens presents a different story. Examples from Tschermak's work, as well as from Weldon's newly found races of commercial peas descended from the highly variable variety known as "Telegraph," show that even for characters which Mendel has studied, such as color or wrinkled seeds, the categories simply are not as robust nor as clear as Mendel would have them. We find a number of different sorts of pea, all of which Mendel would have called (or, perhaps better, would have been forced to call, had he wanted to maintain his division into crisp categories) "smooth," or "yellow" (Fig. 4.1). We also find that the peas that one might call "smooth" if one were habituated to observing a certain variety would have been quite differently described if one had instead chosen a different strain—that is, what passes for "smooth" in one case may well pass for "wrinkled" in another.

Consistent with the general biometrical outlook, Weldon believes that Mendel's greatest mistake is to have ignored the possible influence on an individual's characters arising not only from its parents, but from its ancestry. "Mendel," he writes, "treats such characters…as if the condition of the character in two given parents determined its condition in all their subsequent offspring" (Weldon, 1902a, p. 241). And it is "this neglect of ancestry, the tendency to regard offspring as resembling their parents rather than their race, [which] accounts for much of the apparent inconsistency between the results obtained by different observers who have crossed plants or animals" (Weldon, 1902a, p. 242).

I have chosen to dwell a bit on Weldon's response to Mendel for primarily historiographical reasons. Very often, when the work of the biometrical school is carefully considered at all, it is seen only through the frame of the debate between the

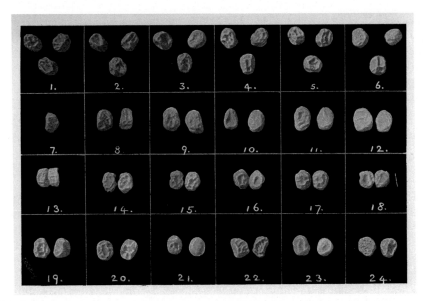

FIG. 4.1

A photograph of peas, included as a plate in Weldon (1902a), showing the smooth gradations that could be found between different character values which Mendel had defined as radically distinct.

Credit: Weldon, W.F.R., 1902. Mendel's laws of alternative inheritance in peas. Biometrika 1 (2), 228–254, plate 1. Oxford University Press. Public domain image.

biometricians and the Mendelians, and this with either a hint of triumphalism in the victory of Mendel, or a wistful nostalgia that the world could have been better had we all just been able to get along a bit sooner, realizing that in the end the two theories could be readily synthesized. While the argument certainly became personally and professionally important for Weldon, Pearson, Bateson, and others, I think it's vital to note that as regards their project for the development of a statistical, mathematical picture of evolution by natural selection, the advent of Mendel simply resulted in more standard biometrical research. Far from being some sort of mythical clash of titans from the very beginning, the affair can be reinterpreted, at least at the scientific level, as a garden-variety dispute over the right way in which to incorporate an interesting collection of data into a preexisting theoretical framework. It is profitable, here, to not let the extremely strong personalities of Bateson, Pearson, and Weldon cloud our vision of the scientific stakes.

Of course, it is also important not to underplay the individual impact or the level of vitriol to which the debate would soon descend. In early June of 1902, Weldon writes to Pearson that he has managed to see an advance copy of Bateson's new book, *Mendel's Principles of Heredity: A Defence*. As Weldon describes it, it consists of "40 small 8vo. pp. of introduction, then a new translation of Mendel, then 100 pp. of demolition of me.—He has really nothing to say; he is simply abusive, and I think his abuse goes rather far beyond permissible limits" (Weldon, 1902b, fol. 1r). Weldon is so hurt by the attack that he asks Pearson to consider removing him from the editorial board of *Biometrika*, so as to not bring the journal into disrepute. If he responds in print, the response should come in *Biometrika* itself given his prior paper on Mendel's results, but, he writes, "the whole thing is paltry and dirty beyond measure, and it is a very serious thing to fill more pages of Biometrika with such stuff" (Weldon, 1902b, fols. 1v–2r).

The conflict with Bateson is, however, nothing if not a spur for Weldon to begin rethinking his position on these issues. He starts by making more precise his concern about Mendelian "characters." He writes to Pearson, complaining that

> *What Bateson does, and what all Mendelians do, is to take the diagram of frequency [Fig. 4.2] and to call a range AB one "character," and the range BC the other "character" of a Mendelian pair.*
>
> *[...]*
>
> *There must be a simple relation between AB, BC, and the [standard deviation] of the original system, which would make the chance that a grandchild falls within BC = ¼?*
>
> *This is not very clearly put...but if one interprets "a chance = ¼" with "neo-Mendelian" latitude, I think it quite likely that a large range of cases of blended inheritance could be shown from the Law of Ancestral Heredity to "obey Mendel's Laws"...*

(Weldon, 1902c, fols. 2r–3v)

Here, then, is the goal Weldon will pursue for the last 4 years of his life. Surely, the argument runs, we can find a way in which the phenomena of Mendelism—a 75%/25% ratio in the grandchild generation—is simply a consequence of some set of parameters for the process of assortment of elements into offspring that Weldon has already contemplated formalizing. The question, then, is what kind of mathematical apparatus would be required to actually describe this process of inheritance, such that we would be able to obtain Mendelian data from one set of initial conditions, and biometrical data from another.

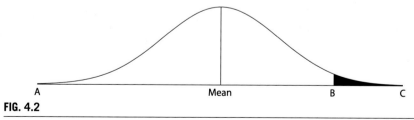

FIG. 4.2

Weldon's normal curve, with "Mendelian" character traits drawn in.

Credit: Drawn by hand by the author, after a letter from Weldon to Pearson, June 23, 1902.

Of elements and chromosomes

While Weldon makes a few false starts at the idea over the course of 1903 and 1904, most of his time is occupied with the by now classic biometrical business of data collection. Multiple genera of plants, birds, and snails make their appearances in the pages of *Biometrika*, along with the beginnings of a protracted experiment with mouse breeding that would, in the end, not be published until years after Weldon's death.

In the first half of 1905, however, Weldon received an invitation to deliver a series of seven lectures on "Theories of the Hereditary Process" at University College, London.[b] The topic of the sixth lecture was to be Mendel's theory, and Weldon laments to Pearson that an illness has rendered him "too stupid to try to get clearer about the Mendelian segregation, and I wanted to do that before talking about Mendel, if possible" (Weldon, 1905h, fol. 1r). Thus, Weldon picks up the question of segregation again with gusto, in hopes of being able to offer his new theory as an element of his upcoming lectures.

In the intervening time since Weldon last prepared a detailed presentation of his thought on this, several things have changed. First, it's clear that by early 1905, he is firmly convinced that chromosomes are the material bearers of what we have been calling the "elements" that are responsible for the production of the characters of organisms. The description of transmission of elements for which he is searching "would start," he writes in a research notebook (to which we will extensively return in a moment), "by taking 'chromomeres' as units," where each of the chromosomes is made up of some static number of chromomeres.

Before continuing, then, we should clarify how it is that Weldon understands the chromosomes, as during this time period there is still extensive debate concerning their nature and function. Unfortunately, Weldon does not offer us any direct citations to source material (having never published any of his discussion of chromosomes during his lifetime), so this reconstruction is somewhat difficult. A letter from Weldon to Pearson in April of 1902 mentions, almost in passing, that "Correns has written a paper to show exactly when the characters in the chromosomes arrange themselves according to Mendel" (see Sloan, 2000, p. 1076; Weldon, 1902c, fol. 2r). While not stated explicitly in the letter, Phillip Sloan's suggestion seems correct that Weldon has just read Carl Correns's article on chromosome theory (Correns, 1902).

Correns had been considering the mode by which the elements (*Anlagen*) would separate themselves in the formation of both normal and germ cells (Rheinberger, 2000, p. 1092). While he only wants to refer to his proposal as a "construction" ("to give something other than a construction…nowadays [is] completely impossible"; Correns, 1902, pp. 305–306), it does still offer us an image of the ways in which elements might be arranged, and one that would be quite compatible with Weldon's later

---

[b] The series of lecture summaries published anonymously in *The Lancet* gives an excellent idea of their content (Weldon, 1905a,b,c,d,e,f,g). I follow the standard citation trend of ascribing them to Weldon, as lecturer; the author of the accounts is, however, unknown (even to *The Lancet*'s editors; pers. comm.).

invocation of chromosomes. Normal somatic cell division as well as normal germ-cell production (which he calls "zygolyte" splitting) are given fairly straightforward explanations:

> We assume that, in the same chromosome, the two elements of each pair of traits lie next to each other (A next to a, B next to b, etc.), and the pairs of elements themselves lie in a row. A picture of this can be found in [Fig. 4.3, 'Fig. 1']. A, B, C, D, E, etc. are the elements of the first parent, a, b, c, d, e, etc. are those of the second parent. In a normal cell- or nucleus-division, which supplies similar products [i.e., similar parent and offspring cells], the longitudinal splitting of the chromosome takes place in such a way that each element is divided in half, in our picture in the plane of the paper. Each half then contains all the elements… In germ cell formation, on the other hand, a longitudinal cleavage takes place once,

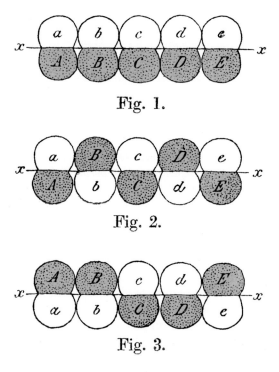

FIG. 4.3

The model of the chromosome proposed by Carl Correns (1902, p. 304). The image labeled "Fig. 1" is the initial orientation of the chromosome, with the elements from each parent appearing down each side of the central axis. In cases of random assortment, however, any set of combinations, such as those in "Fig. 2" and "Fig. 3," could be obtained.

Credit: Rheinberger, H.-J., 2000. Mendelian inheritance in Germany between 1900 and 1910. The case of Carl Correns (1864–1933). C. R. Acad. Sci. III 323 (12), 1089–1096, p. 1092, Fig. 1. Elsevier SAS.

which separates the attachments of the individual pairs...in our image thus per-pendicular *to the plane of the paper, in the line xx ("zygolyte cleavage").*[c]

This is the normal manner in which cell division proceeds. And even in cases where something more complex occurs, we could only detect differences in outcomes (i.e., differences in the daughter cells produced) if there is more than one difference between the elements passed on by the two parents. We do know, however, via experimental work, that what Correns calls "independent assortment" does sometimes happen (i.e., that sometimes somatic and germ cells are produced that are not similar to those of their parents), and that there are indeed cases where there are significant differences between the elements provided by each parent. When this occurs, we must expand the model. At division-time, yet another process must be involved, in which

the elements which come from the same *parent are loosened, A separates from B, B from C, a from b, b from c, etc., but...the two* pairs *of elements are not separated from each other by a longitudinal splitting ("seirolyte cleavage"); the connection of the pairings in the individual pairs remains preserved for the time being (A united with a, B with b, etc.). This loosening enables a change of position of A with a, B with b, etc., independently in each pair from its neighbors and in its outcome dependent on chance. If we stick to our picture, we can assume that the five pairs of elements are rotatable around the axis xx...each receiving an arbitrarily strong (variable) impulse, which sets it into rotation, independently of its neighbors. Then, when calm has returned, 32 different positions will be possible, each of which has the same chances, i.e., will occur the same number of times [e.g., Fig. 4.3, 'Fig. 2' and 'Fig. 3'].*

That is, we could imagine that prior to the somatic or germ cell division at issue, each individual pair of elements can "spin" around the axis *xx*, resulting in a random combination of the elements of the parents. He goes on to propose—to account for cases where, in the very same hybrid, some characters appear to assort independently while some do not—that we might think of this rotation as capable of also stopping after 90 or 270 degrees, leaving, after a separation in the plane of the page, all of the element from one parent in one daughter cell, and all of the element of the other parent in the other daughter cell.

Of course, this is all a fanciful, mechanical "construction"—it's clear that Correns does not intend to propose that chromosomal elements actually spin around their axis!—but it is also clear nonetheless why such a picture could have been compelling to Weldon. First of all, we have a model of chromosome composition that preserves the major points within Galton's work that were attractive to Weldon.

---

[c] The translation is my own, rendering *Anlagen* as "elements"—and with the same caveat for Correns as for Weldon that this is not necessarily a precise terminology for the concept he has in mind.

The chromosomes here play the role of Galton's "stirp," carrying the material basis for heritable traits and placing an emphasis on the order and relative position in which those elements appear, a factor Galton had emphasized that was always particularly significant for Weldon. Further, the process of random assortment that Correns describes is one that would be amenable to the kind of mathematical treatment which Weldon was seeking, and the various possibilities of random and nonrandom assortment that Correns makes room for could allow Weldon the freedom to construct the wide variety of segregation behavior and patterns of inheritance for which he had hoped to account.

Let's see how this all would express itself in Weldon's own words. The clearest presentation comes from a research notebook dated January 2, 1905, entitled "Theory of Inheritance" and headed "About Mitosis" (Weldon, 1905i). We begin with what the chromosomes are not.

> 1.—*It is, I think, evident, from facts of regeneration, that the theory of a nucleus as composed of specific organic determinants is hopeless. It is also evident, from the behaviour of eggs such as those of* Ctenophora *that a special structure, leading to determinate fate of special portions, may exist in [the]* cell-body.
>
> **(Weldon, 1905i, fol. 1r).**

In Weldon's opinion, that is, it is impossible that the nucleus is filled with elements that exclusively give rise to particular character traits, which would "sort out" into the appropriate parts of the organism as it grew. If this were so, then it would be impossible for limbs to be regenerated after they were lost, all the cells carrying the elements producing that limb having been lost as well. That said, the interaction of nuclear elements with the rest of the cell is at this point only dimly understood. Weldon makes reference to phenomena of axial differentiation in the eggs of comb jellies (*Ctenophora*), which seemed to provide decent evidence for the role of not only the contents of the nucleus, but also some further structures in the cells (his waffling here is indicative of serious debate on this matter in the literature at the time). But most importantly of all, Weldon does not take this—or any other recent work on cell biology—to have contradicted in any way his fundamentally Galtonian conception of inheritance:

> 2.—*The above facts do not invalidate [the] conception of nuclear elements as a series of* stirps, *in Galton's sense, each containing something capable of exciting the development of* ~~a whole body~~ *any of the somatic characters, according to its position in the organism.*
>
> 3.—*It seems necessary to regard a* stirp *as capable of exciting, not only somatic characters like those of its parents, but characters like those of its more remote ancestors, under certain circumstances.*
>
> 4.—*It is evident, from the facts of growth and regeneration, that the characters of any one stirp which become active in any one generation are determined by the position of that stirp with reference to the rest—i.e., by a process of the same nature as Mendelian "dominance."*
>
> **(Weldon, 1905i, fol. 1r).**

Here, then, we have Weldon explicitly connecting his understanding of the structure of elements within the gametes to the broader commitments of the biometrical research program. First, the Galtonian emphasis on position and organization. Thanks again in large part to Weldon's insistence on regeneration as an anomalous phenomenon for other conceptions of heredity, we are forced, he argues, to follow Galton in claiming that one of the most significant factors governing which elements within the gametes become patent and which remain latent is the position of the elements with respect to others, both within the cell itself and in neighboring cells. Second, the emphasis on ancestry, which we have already seen him deploy time and again. And lastly, an effort to reconceive dominance as also connected to these very same structural properties of the gametes.

This package of Galtonian beliefs was unwavering for Weldon. In a letter to Pearson a few months before opening this research notebook, he writes that

> *Good old Galton's stirp, in which some of the ancestral characters are latent, is still the only "machine" which will work: and the proper line of research is an enquiry into those embryonic stimuli which make a given character evident or latent. That is my fixed belief.*

(Weldon, 1904, fol. 2r)

This, then, combined with a chromosome-based picture of the elements that make up those stirps, can, Weldon hopes, give us the right ingredients for a mathematical framework describing the transmission of elements from parents to offspring.

Returning to the 1905 research notebook, he writes that if we assume "equality of numbers in germ cells which can fertilize each other…the hybrid zygote should contain 2 sorts of stirps in equal numbers" (Weldon, 1905i, fol. 1r). And more importantly,

> *It seems possible on this assumption to [develop] a theory of nuclear division, which may give Mendel's results without eliminating ancestral influence—i.e., without a theory of "pure" gametes.*

> *Such a theory would start by taking "chromomeres" as units. A chromosome of n chromomeres becomes entangled in the nuclear network: say there are in the zygote $2m$ chromosomes, there will be $2mn$ chromomeres in the resulting nucleus. Before division, these will be gathered into $2m$ groups for an ordinary mitosis, into m groups for a maturation [germ-cell] mitosis…*

(Weldon, 1905i, fols. 1r–1v)

Now we are off to the races. The next 15 pages of Weldon's notebook—as well as several long letters to Pearson and other assorted folios of dense theorizing—are filled with combinatorial arithmetic, trying to work out the various ways in which this process of segregation and grouping might result in different distributions of elements among the offspring within a population. He considers both a variety of sizes of chromosomes and numbers of elements (i.e., values of m and n above), as well as different ways that the chromosomes themselves might be constituted (in essence, versions of Correns's random rotation process, though not in fact that complex).

To enter into the full details of this mathematical treatment would take a turn too technical for my goals here.[d] But it is important to consider Weldon's results—or lack thereof. Recall the very reason that he sets out on this project, discussed in his 1902 letter to Pearson, and reemphasized in the 1905 research notebook. Weldon is hoping that, armed with a mathematically analyzable theory of the distribution of elements within chromosomes, he has some chance of constructing a theory of inheritance that could both do justice to biometrical results on blending inheritance and derive Mendelian outcomes as a special case—only by describing the way in which the elements segregate and associate from parents to offspring.

Unfortunately, the mathematics simply don't work. As he writes to Pearson concerning his first effort (i.e., the first of his sets of assumptions about chromosome size and formation),

> I have laboriously worried through the effect of supposing the chromosomes to retain their individual constitution right on from the moment of fertilization to the formation of new germ-cells; and it does not give anything like a proper segregation: there are very few "pure" individuals, among either dominants or "recessives."

> **(Weldon, 1905j, fol. 1r)**

That is, in this and, eventually, every way that Weldon can think of to formalize the process of elements forming chromosomes and assorting into offspring, it simply refuses to produce a Mendelian 75%/25% proportion.[e] Weldon's initial effort is a failure; there will be no general theory of inheritance built on these grounds.

At the same time, he is fairly convinced that he knows exactly what has gone wrong. As he continues in the above letter to Pearson,

> Of course, the thing really wants more than this. It wants segregation of elements, but what I have not and cannot put into it is the view of each element as a "stirp."—Dominance of yellow seed-colour in a Pea is not merely prevalence of the effect, due to one set of elements, over that due to the other set in the hybrid: it is also an excitation of the ancestral yellow-seededness of the race of green peas used in making the cross.

> **(Weldon, 1905j, fols. 1r–1v)**

The model still lacks what is, for Weldon, a key Galtonian insight: that the probability of an element's being latent in a future generation is not merely a result of the number of those elements found in a nucleus, but also depends upon whether or not that element was active in previous generations. As a particular element becomes "excited" or "activated," this is in turn likely to cause that element to be active in

---

[d] Readers interested in the mathematical complexities can find my reconstruction of all of Weldon's arithmetic in the scholarly edition of Weldon's *Theory of Inheritance* manuscript (Radick et al., in prep).

[e] Matters are even worse, in fact; as I have interpreted Weldon's mathematics, the three cases which he painstakingly considers and hopes will provide him significantly different outcomes differ only by 1.3% in their ratio of dominant offspring, all around 65%.

immediate future generations as well. The very next day, he writes a large-format, five-page letter to Pearson that triumphally begins:

> *Here is the true gospel, or a sort of approximation to it, at last! When a stirp goes into a zygote, it carries a lot of properties, but those which are manifested by the body into which the zygote develops are transmitted with increased intensity to the gametes of that body, thus establishing that correlation between character of parent and character of its reproductive cells, which I had foolishly been unable to put in. But if a stirp, having become active in this way, be introduced into a zygote in which the majority of stirps are so active in other directions that its own properties become latent in the body to which the zygote gives rise, then that stirp transmits its properties in a weakened condition to the next generation.*
>
> *If you apply this luminous principle to Peas, you get Mendel pat.*
>
> **(Weldon, 1905k, fol. 1r)**

Weldon goes on to note that, in order to figure out how to connect his assortment and segregation work with this new insight, he needs a way to assign a new property to each element—he calls it "valency"—that describes the intensity of this "activation" over time. This, in turn, would have to be related to the statistical properties of the distributions of elements produced by his theory of segregation. But that would require him to be able to calculate facts about the distribution that he does not yet know how to derive, in particular, its probable error. "You see," he writes to Pearson, "I can't even find the [standard deviation], and if I could, I could certainly not find the relation between it and the probable error!" (Weldon, 1905k, fols. 3r–3v). All he can hope to do, he concludes, is hunt for approximations.

But at this point, Weldon is out of time—he has to deliver his lectures on Mendel in January and February, and so he sets aside his theoretical work. Over the course of late 1905 and early 1906, he begins to fashion the material that he had collected for the lectures into a book-length treatment of inheritance. In the last extant draft of its second chapter, we have a promissory note that he will return to his chromosome work later in the volume. "We shall see," he writes, "in a future chapter how far the suggestion made [by Galton] can be brought into harmony with the facts of germinal structure which have been discovered during the last thirty years" (Weldon, 1905l, Chapter 2, p. 2, fol. 67r; see also Radick et al., in prep).

Unfortunately, that chapter was never written. Weldon, habitually overworked and never in excellent health, took the long trip from Oxford to London on Tuesday, April 10, 1906, in spite of a bout of influenza, visited a gallery on Wednesday, and the dentist on Thursday. He was taken directly from the dentist to the doctor, and was dead of pneumonia by Friday—Good Friday, April 13, at the age of 46.

Pearson was crushed. "It seemed possible to go on," he wrote in a letter to Galton at the end of April, "so long as I was attempting to place things at Oxford in order, but I am quite dazed & for the first time I think in all my teaching work, the idea of facing my students and lecturing seems positively repellent, at times impossible. I feel simply without energy to meet the term" (Pearson, 1906b, fols. 2–3; Fig. 4.4).

FIG. 4.4

Karl Pearson, 1910.

Credit: Pearson, E.S., 1938. Karl Pearson: an appreciation of some aspects of his life and work. Part II: 1906–1936. Biometrika 29 (3/4), 161–248, 161–162.

Taking stock

Over these last two chapters, we have seen the broad outlines of the biometrical program, the ways in which they had hoped—hopes that were only partially realized—to construct a mathematical, a statistical, a chancy theory that could fully account for the breadth of phenomena that Darwin, and later Mendel, had described. It is time to step back, then, and consider the philosophical and biological commitments that enabled this shift, and that made it appear to Pearson, Weldon, and a number of their students (the first to have enough training in both statistics and biology to be qualified to evaluate their combination) the most likely way in which we could develop a complete picture of Darwin's gradualist, incremental theory of evolution by natural selection.

That they were fully committed to doing so is evident—even in the face of significant pressure from a wide array of their own colleagues. As Weldon described it, "this contention that numbers 'mean nothing' and 'do not exist in Nature,' and so on, is a very serious thing, which will have to be fought. Most other people have got beyond it, but most biologists have not" (Weldon, 1900c, fol. 1v). We thus should aim to discover what it was that made the biometrical world view so distinctive.

First of all, it is important to see the precise sort of "population thinking," to borrow Ernst Mayr's later and much-misused phrase, that was at work in the biometricians. As Margaret Morrison (2002) has noted, it is of an interestingly different sort than that which we will find in Chapter 6, in the later works of R. A. Fisher.

What we have here are populations of actual, real-world organisms—nothing like an infinite gene pool or the other kinds of mathematical abstractions that will soon take up residence at the core of Darwinian theory. Galton, for instance, often considered the normal curve as applying to, say, the statures of actual samples from the population of England, forming curves within which each person would have their designated place (Galton, 1892, pp. xi–xii). Such a curve describes real people from the real nation of England, not some mathematical idealization thereof. And it is this power to describe the world around us that Weldon, at least, took to be the most important reason to turn to statistical methods. "It is the first business of a scientific man," he writes in a contribution to a volume on the methodology of science (written late in 1905), "to describe some portion of human experience as exactly as possible. It does not matter in the least what kind of experience he chooses to collect; his first business is to describe it" (Weldon, 1906, p. 81).

Biologists, as the biometricians would have it, have been at a deep and abiding disadvantage when it comes to their efforts to describe their experience. Consider one of Weldon's favorite comparisons: that with Lord Rayleigh's results that led to the discovery of argon. We start with a contrast between two ways in which one could prepare samples of nitrogen: either by isolation of the gas from a sample collected from the atmosphere (removing the oxygen, carbon dioxide, water vapor, and so forth, using known chemical processes), or by the creation of a new sample through performing chemical reactions which were known to give off pure nitrogen gas (such as breaking down an ore which contained nitrogen). Rayleigh created what should have been equal masses of nitrogen using the two different methods, but when he weighed them, he found a persistent, irreducible difference of around 0.01 g. What makes the case special is the further fact that the methods of physicists are sufficiently precise to justify the following inference: this was *not* an experimental mistake, an accident, or an expected error in the use of our measurement apparatus. Rather, we can conclude that there is a real phenomenon here, and, when that phenomenon was pursued in greater detail, the discovery of argon was the result. In Weldon's words, the most important step in that process was the "refusal to replace the variable and discordant experience of the weight of nitrogen by an ideal uniformity based on the mean of the actual experiences" (Weldon, 1906, p. 92).

The extent to which this can be sharpened into a critique of the then-current practice of biology is not left as an exercise for the reader.[f] Physical scientists have "already succeeded in confining the limits within which these inconsistencies [in measurement, experimental error, etc.] occur, so that the proportion of the overall experience affected by them is very small. But biologists have not yet advanced so far as this: the margin of uncertainty in their experience is still so large that they are obliged to take account of it in every statement they make" (Weldon, 1906, p. 93). Because biologists cannot justifiably neglect the error implicated in any biological claim, it is quite simply intellectually dishonest to present those results without

---

[f] I have argued elsewhere that it also constitutes a criticism of the austere positivism of Pearson (Pence, 2011).

representing for the reader, in some clear and straightforward way, the full array of data, rather than a summarized, averaged, or abstracted value. Of course, it will by now be no shock to learn that the easiest, if not the only, way to do this is via statistics. The explicit claim as Weldon phrases it is rather anodyne, but the implicit critique is scathing: all biological results presented without taking into account the full range of their distribution are suspect. This commitment to statistical description of real populations is the first tenet of Weldon's philosophical approach.

But it leaves us with an obvious further question: why is it that biologists cannot control their errors in the way that their colleagues in the physical sciences have mastered? Is this an endemic problem to the field, to the subject matter itself, or is it rather something that could, at least potentially, be overcome with further work, higher quality measurements, better theories, and so on? Weldon is not particularly clear here—in part, he was still coming to terms with these ideas himself, having been forced to reconsider his approach to chance and statistics while sparring with Bateson and working on his chromosome theory. At times, he sounds as though he is offering a straightforward enough endorsement of an ignorance interpretation of the necessity of statistics. "All experience," he writes, "which we are obliged to deal with statistically, is experience of results which depend upon a great number of complicated conditions, so many and so difficult to observe that we cannot tell in any one case what their effect will be" (Weldon, 1906, p. 97). Because, for instance, we cannot know the myriad small influences—lengths of bones, size of cartilage, flexibility of tendon, and so on—that go into the determination of the height of a person, all we can do is measure the height of their father and mother, thereby learning at least one important condition on which the whole complex mess rests, and using it to make, via regression, the best kind of inference of which we are technically capable.

Such a position, when Weldon presents it in this way (Weldon, 1895, or in the conclusions of his crab papers; Weldon, 1893), seems to rely on the idea that at least part of the utility of statistics derives from its ability to provide us best-case predictions in environments of incomplete or poor data, or unknown underlying causes. Thus, the second tenet of Weldonian philosophy of biology: in addition to offering us the best known way to describe all of our biological experience, errors included, we also must introduce statistics into biology as a vitally important tool to enable predictions in our usual working conditions, given these impediments. Note that, consistent with the first tenet, these remain predictions from actual, concrete populations to future generations of those populations. As Weldon puts the matter in a letter to Pearson, "one has a chance of predicting *in a given case, from a knowledge of parental variability,* what will happen" (Weldon, 1902c, fol. 3v).

But this can only be part of the story. At the same time, we must square this obvious theme in Weldon's thought with his program of chromosome research. Weldon certainly also believed that there would someday be a way in which we could elaborate a causal (one is tempted to say machine-mechanical, in the sense of Nicholson, 2012) picture of the relationship between the elements of parental chromosomes and the elements of offspring chromosomes. Such a relationship could be generalized by aggregating it statistically, but Weldon sees that such an effort will need nonetheless to take account of the mechanisms of segregation, assortment, and "valency" as traceable through cell divisions in particular individuals leading to particular offspring.

Here, then, I think we can detect another, more subtle point arising from Weldon's analysis of the differences between physical and biological sciences mentioned above. There are two kinds of conclusion one might draw from Weldon's discussion of statistics as a way to preserve and present error in the biological context. One we have already seen: the failure of non-statistical approaches to represent the phenomena of inheritance and evolution. But another, equally pressing obligation is to resolve the underlying situation itself—to render biological science better able to make the kinds of precise predictions that permitted Rayleigh to discover argon in the first place. One obvious and compelling way in which we might do this, then, would be to better elucidate the laws of inheritance and the connections between those laws and broader population outcomes. Thus a third tenet of Weldon's philosophy of biology: seek to ground statistically expressed trends in their underlying causal laws whenever possible.

Given the historical arc of the biometrical program, I suspect that it is this last approach that would have taken more of Weldon's focus, had he lived. His increased emphasis on natural selection—recall the silting of Plymouth Sound and Weldon's search for the reason that decreased frontal breadth in his crabs could have been favored by natural selection—as well as his increased interaction with cell biology and study of chromosomes in an effort to effectively understand Mendelism on his own terms, both seem to constitute impulses in this direction.

In any event, we see a clear and consistent philosophical approach to the statistical study of evolution. Biological systems are complex—too complex to be apprehended without significant error—and hence we must use statistical methods to describe them, both because these let us be honest about such error, and because they let us make positive predictions, to the extent of our ability. But we must be equally devoted to the discovery of the fundamentals of such systems. A philosophical approach every bit as sophisticated as we might have expected from a biologist both as innovative and as self-reflective as Weldon.

Whither biometry?

It is something of a historical commonplace that Weldon's death concludes the conflict between biometricians and Mendelians, Pearson retreating largely into his eugenics work and ceasing to contribute to the biometricians' program for evolutionary theory in any real way. William Provine, for instance, writes that Pearson "still published an occasional criticism of Mendelian interpretations, but he did not want to engage again in controversy. So in England the conflict died away" (Provine, 1971, p. 88). On such an interpretation, then, whatever the virtues of the biometrical approach, we must wait for the brilliance of Fisher, Wright, and the other architects of the Modern Synthesis to rebuild the foundations of a statistical biology—largely without utilizing any of the theoretical work of the biometricians.

If this is true, one would be forgiven for wondering why we have spent the last three chapters detailing the efforts of the biometrical school. Of course, they might have intrinsic interest for those with a taste for historical minutiae, but they would not then have any real impact on the later development of the biological sciences. My aim in the last part of the book, then, is to argue that we must, indeed, understand the work of the biometricians in order to properly appreciate the philosophical foundations of the use of chance and statistics in the Modern Synthesis and beyond. To an extent that has not been sufficiently appreciated by historians of biology (and contra the explicit assertions of those like Provine), there is real continuity here, connecting the work of Weldon and Pearson, via the Synthesis, to contemporary biology—there is, to put it differently, a very good reason to take the insights of Weldon and Pearson seriously. First, however, we must make an intermediate stop. Evolutionary theory did not freeze between Weldon and Fisher.

References

Bateson, W., 1901. Heredity, differentiation, and other conceptions of biology: a consideration of Professor Karl Pearson's paper 'On the principle of homotyposis'. Proc. R. Soc. Lond. 69, 193–205. https://doi.org/10.1098/rspl.1901.0099.

Bateson, W., 1902. Mendel's Principles of Heredity: A Defence: With a Translation of Mendel's Original Papers on Hybridisation. Cambridge University Press, Cambridge.

Correns, C., 1902. Ueber den Modus und den Zeitpunkt der Spaltung der Anlagen bei den Bastarden vom Erbsen-Typus. In: Gesammelte Abhandlungen zur Vererbungswissenschaft aus periodischen Schriften 1899–1924. Springer-Verlag, Berlin, pp. 300–314.

Galton, F., 1892. Hereditary Genius: An Inquiry Into Its Laws and Consequences, second ed. Macmillan, London.

Morrison, M., 2002. Modelling populations: Pearson and Fisher on Mendelism and biometry. Br. J. Philos. Sci. 53, 39–68. https://doi.org/10.1093/bjps/53.1.39.

Nicholson, D.J., 2012. The concept of mechanism in biology. Stud. Hist. Phil. Biol. Biomed. Sci. 43, 152–163. https://doi.org/10.1016/j.shpsc.2011.05.014.

Pearson, K., 1899. Letter from KP to WFRW, 1899-02-28.

Pearson, K., 1906a. Letter from KP to FG, 1906-05-13.

Pearson, K., 1906b. Letter from KP to FG, 1906-04-29.

Pence, C.H., 2011. "Describing our whole experience": the statistical philosophies of W. F. R. Weldon and Karl Pearson. Stud. Hist. Philos. Biol. Biomed. Sci. 42, 475–485. https://doi.org/10.1016/j.shpsc.2011.07.011.

Provine, W.B., 1971. The Origins of Theoretical Population Genetics. Princeton University Press, Princeton, NJ.

Radick, G., 2011. Physics in the Galtonian sciences of heredity. Stud. Hist. Phil. Biol. Biomed. Sci. 42, 129–138. https://doi.org/10.1016/j.shpsc.2010.11.019.

Radick, G., 2015. Beyond the "Mendel-Fisher controversy". Science 350, 159–160. https://doi.org/10.1126/science.aab3846.

Radick, G., Pence, C.H., Shan, Y., Jamieson, A., in prep. W. F. R. Weldon's Theory of Inheritance.

Rheinberger, H.-J., 2000. Mendelian inheritance in Germany between 1900 and 1910. The case of Carl Correns (1864–1933). C. R. Acad. Sci. III 323, 1089–1096. https://doi.org/10.1016/S0764-4469(00)01267-1.

Sloan, P.R., 2000. Mach's phenomenalism and the British reception of Mendelism. C. R. Acad. Sci. III 323, 1069–1079. https://doi.org/10.1016/S0764-4469(00)01255-5.

Weldon, W.F.R., 1893. On certain correlated variations in *Carcinus mœnas*. Proc. R. Soc. Lond. 54, 318–329. https://doi.org/10.1098/rspl.1893.0078.

Weldon, W.F.R., 1895. An attempt to measure the death-rate due to the selective destruction of *Carcinus mœnas* with respect to a particular dimension. Proc. R. Soc. Lond. 57, 360–379. https://doi.org/10.1098/rspl.1894.0165.

Weldon, W.F.R., 1900a. Letter from WFRW to KP, 1900-10-16.

Weldon, W.F.R., 1900b. Letter from WFRW to KP, 1900-12-03.

Weldon, W.F.R., 1900c. Letter from WFRW to KP, 1900-11-16.

Weldon, W.F.R., 1900d. Letter from WFRW to KP, 1900-12-12.

Weldon, W.F.R., 1901. Letter from WFRW to KP, 1901-05-22.

Weldon, W.F.R., 1902a. Mendel's laws of alternative inheritance in peas. Biometrika 1, 228–254. https://doi.org/10.1093/biomet/1.2.228.

Weldon, W.F.R., 1902b. Letter from WFRW to KP, 1902-06-03.

Weldon, W.F.R., 1902c. Letter from WFRW to KP, 1902-06-23.

Weldon, W.F.R., 1904. Letter from WFRW to KP, 1904-10.

Weldon, W.F.R., 1905a. Current theories of the hereditary process [1]. Lancet 165, 42.

Weldon, W.F.R., 1905b. Current theories of the hereditary process [2]. Lancet 165, 180. https://doi.org/10.1016/S0140-6736(00)94753-6.

Weldon, W.F.R., 1905c. Current theories of the hereditary process [3]. Lancet 165, 307–308. https://doi.org/10.1016/S0140-6736(01)42847-9.

Weldon, W.F.R., 1905d. Current theories of the hereditary process [4]. Lancet 165, 512. https://doi.org/10.1016/S0140-6736(01)44351-0.

Weldon, W.F.R., 1905e. Current theories of the hereditary process [5]. Lancet 165, 584–585. https://doi.org/10.1016/S0140-6736(00)78433-9.

Weldon, W.F.R., 1905f. Current theories of the hereditary process [6]. Lancet 165, 657. https://doi.org/10.1016/S0140-6736(01)45697-2.

Weldon, W.F.R., 1905g. Current theories of the hereditary process [7]. Lancet 165, 732. https://doi.org/10.1016/S0140-6736(01)45769-2.

Weldon, W.F.R., 1905h. Letter from WFRW to KP, 1905-01-02.

Weldon, W.F.R., 1905i. MS notebook of Weldon's entitled "MCMV. Theory of Inheritance" dated 2 Jan 1905.

Weldon, W.F.R., 1905j. Letter from WFRW to KP, 1905-01-11.

Weldon, W.F.R., 1905k. Letter from WFRW to KP, 1905-01-12.

Weldon, W.F.R., 1905l. MS of Theory of Inheritance Book.

Weldon, W.F.R., 1906. Inheritance in animals and plants. In: Strong, T.B. (Ed.), Lectures on the Method of Science. Clarendon Press, Oxford, pp. 81–109.

Weldon, W.F.R., Pearson, K., Davenport, C.B., 1901. Editorial: the scope of Biometrika. Biometrika 1, 1–2. https://doi.org/10.1093/biomet/1.1.1.

Reconciling the biometrical conclusions: Evolution from 1906 to 1918

5

While it is a confusion of thought to oppose a statistical conclusion and a physiological interpretation, it cannot be denied that the Galtonian and Mendelian views of heredity are not yet in harmony.
J. Arthur Thomson, *Heredity* (1908)

The "interregnum" period between the death of Weldon in 1906 and the arrival of the Modern Synthesis with the now-heralded works of Fisher, Wright, and others over the late 1910s and 1920s is often passed over in relative silence. As the common history would have it, Bateson and the other Mendelians won the fight that unfolded across the last chapter—or, at the very least, the biometricians ceded the field of battle. More importantly, none of it mattered anyway, because R. A. Fisher would derive the entire statistical formalism of evolutionary theory completely on his own, from scratch, largely as a result of his early training in, and then reliance on, statistical physics. The period from 1906 to 1918, then, is a sort of late-Mendelian wasteland, not worth our time, except insofar as it harbors a debate of minor interest on the question of whether natural selection acts on continuous or discontinuous variations (such a view can be found, e.g., in Provine, 1971). If anything, the attitude most often taken toward biometry is a kind of vague regret. The geneticist Alfred Henry Sturtevant, for instance, writes that the "personal quarrel" between Bateson and Weldon "certainly delayed the utilization of the powerful methods of statistics in much of genetics" (Sturtevant, 2001, p. 58). The possibility of synthesis, on this view, should have been obvious, had personal animosity not clouded the vision of the biologists in a position to see it. There are a number of falsehoods in this traditional history, and unpacking them—and the philosophical consequences which result—will be my project in what remains of the book.

Before we continue, however, we should clarify our target. The thread that I have been tracing in the first four chapters, detailing the introduction of concepts of chance and statistical theorizing into evolutionary theory, will become rather diffuse over this decade, and so it's important, I think, to ensure that we know what exactly it is that we're searching for. How, in particular, should we understand the late-biometrical (or Weldonian) perspective in the years between Weldon's death and the advent of the Modern Synthesis?

Generalizing a bit from the conclusion to the last chapter, I want to emphasize four aspects of the biometrical view that will go on to be vitally important to

The Rise of Chance in Evolutionary Theory. https://doi.org/10.1016/B978-0-323-91291-4.00002-9

contemporary biology, three of which are commonly taken to be absent (or, if present, neglected by mainstream Mendelians) until Fisher "rediscovers" them. At the end of his life, we saw that Weldon was looking for a *statistical* theory of evolution, which could let us mathematically understand the *action of natural selection* across generations, at the *population level*, and which could be harmonized with *Mendelian transmission*, at least as a special case. I won't elaborate on any of those four elements as they appeared in biometrical work any further—the last two chapters provide ample demonstration of what Pearson and Weldon, at least, meant by each of the four and their combination—but laying them out clearly suffices for establishing my argument. While it indeed is the case that Weldon was (save one more author whom I will discuss below) the last biologist before Fisher to explicitly work toward a version of evolutionary theory that would satisfy all four of these desiderata, none of the four fell out of repute in the period between Weldon's death and the early works of the Modern Synthesis. Setting aside cases of personal antagonism, there was little expression or discussion of incompatibility between any of these four goals for evolutionary theory, and much work was performed, on both sides of the Atlantic, with the aim of discovering ways of presenting evolution that could combine pairs and triplets of them in various novel and interesting ways. In short, the course of evolutionary theory between 1906 and 1918 looks much more like the collaborative, piecemeal development of an integrated picture of evolution than an exclusively Mendelian world following the complete defeat of the biometricians. The rumors of the death of biometry have been greatly exaggerated.

Where not to look

We should begin by considering two places in which it is *not* useful to look for evidence of this sort of collaborative, continuous theory building. One of the few caveats in my thesis above noted that we ought to set aside cases of personal antagonism. Indeed, the acrimony that had developed between Karl Pearson and William Bateson was much too deep to be replaced by anything like cooperation, even after the shock of Weldon's untimely death. Given their importance, however, both to my story here and to the history of biology more generally, I want to begin by gesturing at a few final conflicts. They continued to offer much heat and little light.

The late Pearson

Weldon's passing would, of course, mark a significant moment in the life of his closest collaborator. Without Weldon by his side to engage in skirmishes in defense of biometry against the Mendelians, Pearson came to be less inclined to participate in the venues where such debates were fiercest. After one of the rare instances in which he did so—a series of biometry-Mendelism speeches arranged at the Royal Society of Medicine in late 1908, primarily pitting Pearson against Bateson's assistant G. P. Mudge—he wrote to Galton, lamenting that

I am not a ready debater & find it hard to marshal my arguments in reply to a set speech of nearly 70 minutes designed to prove that biometry was shear [sic] rubbish & medical men would be fools to give any help to a biometrician. It is on these occasions I miss so much Weldon's ready repartee & light cavalry charges into the foe! I don't know how far I saved defeat; if I did at all it was owing to the unmeasured abuse of my opponent.

(Pearson, 1908, fol. 2)

Those familiar with Pearson's own famously acerbic personality may be surprised to hear him confess such a weakness, but it seems as though the absence of Weldon's personal blend of encyclopedic knowledge of biological facts and eagerness to engage with the Mendelians led to Pearson's disenchantment with the conflict.

This also coincided with the pressure of institutional factors. Pearson's Biometric Laboratory had, since securing unencumbered access to its funding in 1903, taken up a larger and larger share of his time. To this was added in 1906 the directorship of the Francis Galton Eugenics Laboratory, formerly the Eugenics Record Office, now also expanded with yet more financing. Pearson would soon have a significant staff at both institutions, as well as continued (effectively sole) editorship of *Biometrika*. His approach to statistical science never changed, consisting to the end primarily in the steady, cautious collection of reams of data, much of it published either in *Biometrika* or directly in the in-house publication series of the Eugenics Laboratory.

In short, Pearson had little time to fight these battles, whether in the correspondence pages or in person. Setting aside two further exceptions to which we will turn below in discussing William E. Castle and G. Udny Yule, Pearson after Weldon's death wrote only four papers on the theoretical basis of evolution. Three came in a quick flurry in 1909 and 1910, spurred largely by attacks from Davenport and Bateson on Pearson's formulation of the law of ancestral heredity (Pearson, 1910, 1909a, b). These are quite narrowly focused, designed really only to show that particular, technical arguments made by opponents had already been responded to in other work. Finally, there is one peculiar article, published in 1930, entitled "On a new theory of progressive evolution" (Pearson, 1930). This paper is interesting primarily because it demonstrates the extent to which Pearson has lost touch with the frontier of biological research. By 1930, notably, we have already seen the development of a fair bit of the theoretical apparatus of the early Synthesis, as well as its significant integration with the study of cellular biology and chromosomes by then well underway in the laboratory of Thomas Hunt Morgan and elsewhere. Pearson does not mention any of these results. Rather, he demonstrates, in essence, that if the form of individual variability—the distribution of individual variation around the parental mean—remains constant, then inbreeding of unusual parents could, by biasing the hereditary process toward the extreme ends of the spectrum, drive a "progressive" trend in evolution. Such progress could, he writes, continue at least until the rest of the organism could no longer bear further modification of a given trait without harm to the overall function of the system as a whole. He cites no theoretical work but his own, and nothing more recent than the 1909 papers mentioned above. He has, in essence, voluntarily left the mainstream of biological research.

The late Bateson

Bateson, for his part, never understood the utility of mathematical inquiry for biological research, and was thus increasingly left behind as even developments within Mendelism became steadily more mathematical. While many of his late publications could be cited as proof of the point, it is perhaps best to consider his textbook, *Mendel's Principles of Heredity*,[a] published in 1909 and aiming to offer what he considered "a useful and relatively permanent presentation of the phenomena" of Mendelian inheritance (Bateson, 1909, p. vi). At this point, following the common view, one would expect Bateson to be exulting in victory—and indeed, that is exactly what one finds, with extra helpings of sneer. In the first chapter, he writes that

> *Of the so-called investigations of heredity pursued by extensions of Galton's non-analytical method and promoted by Professor Pearson and the English Biometrical school it is now scarcely necessary to speak. That such work may ultimately contribute to the development of statistical theory cannot be denied, but as applied to the problems of heredity the effort has resulted only in the concealment of that order which it was ostensibly undertaken to reveal. [...] To those who hereafter may study this episode in the history of biological science it will appear inexplicable that work so unsound in construction should have been respectfully received by the scientific world.*

(Bateson, 1909, pp. 6–7)

Little surprise, then that there is no move toward synthesis here. We can, however, gain an idea as to possible explanations for the construction of our traditional history. For Bateson and Pearson both, the fight between biometry and Mendelism is clearly the defining feature of biology during this period. For both men as well, biometry has been defeated, and Mendelism has won. Neither Pearson's lamentations nor Bateson's gloating will much advance the science of biology, and so we must wait for later impetus, unclouded by the persistent animus between the two men. Without a wider lens on biological practice during this period, it is difficult to see any other way to analyze the situation.

Beyond gloating, however, Bateson also bitterly resisted the drive toward the mathematization of Mendelian theory, as well as the application of Mendelian insight to the construction of new approaches to evolution, as opposed to individual inheritance. Rather than exploring the ways in which Mendelism might help us understand the effect of natural selection, he writes, "it is as directly contributing to the advancement of pure physiological science that genetics can present the strongest claim. We have an eye always on the evolution-problem. We know that the facts we are collecting will help in its solution; but for a period we shall perhaps do well to direct our search more especially to the immediate problems of genetic physiology" (Bateson, 1909, p. 4). Better, that is, to explore the question of the physiological basis

---

[a] Not to be confused with his *Mendel's Principles of Heredity: A Defence*, published in 1902 amid the conflict with Weldon and discussed in the last chapter.

of Mendelian characters—believed by many at this point, but not Bateson, to reside in the chromosomes (Cock, 1983)—and leave the explanation of natural selection for future scientists.

When it came to the usefulness of statistics, however, Bateson is less conciliatory, aiming more directly to pass his hatred for biometrical methods on to other researchers. "Professor Pearson and others committed to these methods," he writes, "have of late defended their position by arguing that there is no fundamental incompatibility between Laws of Ancestral Heredity and the conclusions of Mendelian analysis. The matter would not be worth notice were it not that the same proposition is being freely repeated by several writers seeking some convenient shelter of neutrality" (Bateson, 1909, p. 130). By 1909, however, Bateson's work reveals that he, too, has essentially left the mainstream of biological research. Taking little heed of his admonition, researchers around the world had already begun to consider the application of statistical methods both to problems of inheritance and to a Mendelian take on natural selection, a trend that would only accelerate in the years following the publication of Bateson's textbook.

Statistics without a statistical theory of inheritance

With Pearson and Bateson set aside, then, we can return to the various ways in which the four central threads of biometrical research that I laid out above were advanced over the course of this decade. Let's begin with statistical methodology. Of course, as I have repeatedly noted, I have no ambitions here to tell a history of statistical methods in and of themselves. The work catalyzed by the efforts of Pearson would very quickly produce a massive efflorescence of statistical research, and by just a few decades later, this would yield the fundamentals of much of today's mathematical statistics. That, however, would be the topic for a different book (see, e.g., Gigerenzer et al., 1989; MacKenzie, 1981; Stigler, 1986). For my purposes, on the contrary, I want to consider a different question: how were these methods being adopted by biologists—even by biologists who were ambivalent at best about the utility of biometrical theory more generally?

The first way in which statistical methods began to see wide adoption mirrors their most common use today: in data analysis. Regardless of whether such an approach is integrated into one's understanding of biological theory, and even in cases where the methods of statistics were recruited to support "purely Mendelian" conclusions, they were quickly seen to be essential for the understanding of broad biological experiments of the sort that were increasingly undertaken during this time. As researchers (often allied to agricultural stations) performed more and larger breeding experiments with the aim of better understanding Mendelism, statistical methods took a correspondingly greater role as they became necessary for grappling with the quantities and types of data those experiments generated.

Let's consider two examples, one from each side of the Atlantic. William E. Castle (Fig. 5.1) studied under Charles B. Davenport—the third, with Pearson and Weldon,

FIG. 5.1

William Ernest Castle.

Credit: Snell, G.D., Reed, S., 1993. William Ernest Castle, Pioneer Mammalian Geneticist.
Genetics 133 (4)751–753, p. 751.

of the founding editors of *Biometrika*, but a later convert to Mendelism—at Harvard. He became something of a partisan of Bateson's early in his career, engaging in some fruitless bickering with Pearson over a mistaken reading of the Law of Ancestral Heredity in the years just after 1900 (Castle, 1903; Pearson, 1904a; Pearson returned to the same subject much later in Pearson, 1911), and he wrote a number of articles pushing for Bateson's view of natural selection as proceeding by discontinuous mutation (see Provine, 1971, pp. 108–114). This meant—as it did for many in the period—that Castle would no longer have a friendly relationship with Pearson. Castle would soon, however, begin a massive breeding project with his student Hansford MacCurdy on rats and guinea pigs, hoping to clarify the patterns of transmission in their coat colors, an arena in which a variety of apparently contradictory results had been obtained in recent years. Weldon himself, building on the work of his collaborator Arthur Darbishire (see Ankeny, 2000), had studied the same subject in the years before his death, and it was thought that the data could form a crucial test of the limits of natural selection's action in a Mendelian context (Weldon's few surviving results may be found in Pearson, 1907a). For Castle's part, he would emerge from the episode a supporter both of the importance of natural selection and of the utility of statistical methods for the analysis of data.

That data would run to thousands of observations of rats and guinea pigs, and MacCurdy and Castle showed no hesitation in presenting their results statistically. After offering some brief, estimated summary interpretations of this material, they write that

> *For those, however, who place confidence in the more precise methods of statistical analysis devised by Pearson and others, it may be more satisfactory to treat the tables, which have been constructed for the various groups of individuals, as* correlation tables, *and derive from them the constants which measure the variability of parents and children respectively in the several groups, and the degree of correlation between the two.*
>
> **(MacCurdy and Castle, 1907, p. 25)**

Indeed, this is exactly what they have done, deriving a few standard correlation measures between generations in around 1000 of their guinea pigs. The correlations show, further, that directional change is indeed possible in a character trait that is governed by Mendelian patterns of transmission. In their own words, while color patterns "are mutually alternative in heredity," which is to say that they "in general do not overlap; they form a discontinuous series," it is nonetheless the case that

> *these types may be modified in two different ways: (a) By selection of abmodal variates within the same continuous series, and (b) by cross-breeding between different types. There is no evidence that one of these methods has effects less permanent than the other. So far, then, as these experiments go, they support the Darwinian [continuous] view rather than that of De Vries [discontinuous].*
>
> **(MacCurdy and Castle, 1907, pp. 3–4)**

Modification of a type by selection *within* a continuous series, that is, proves for MacCurdy and Castle to be equally effective as modifying types by hybridization *across* those same continuous series. And statistical tools, as a method for the manipulation of the data needed to draw that conclusion, are second nature, even to a convinced Mendelian and erstwhile opponent of Pearson.

In his opinionated work on the history of genetics, Sturtevant, one of the key early figures in T. H. Morgan's work on the genetics of *Drosophila*, lavishes the following praise on our second example of the heavy use of statistics outside biometry:

> *But the most important influence leading to the general use of statistical methods was that of the Danish botanist [Wilhelm Johannsen], beginning in 1903 and culminating in his* Elemente der exakten Erblichkeitslehre *(1909). Johannsen... pointed out that the results of the biometrical school were only valid statistically, were of no help in individual families, and gave no insight into the mechanisms involved. But he did recognize the value of statistical methods, and used them extensively.*
>
> **(Sturtevant, 2001, p. 59)**

Wilhelm Johannsen (Fig. 5.2) is today best known for the development of his "pure line theory." He isolated cultures of beans which had self-fertilized for a large number of generations, and in which, upon further cultivation, he saw precisely no heritability of individual variations—offspring regressed entirely to the mean of the pure line itself, making evolution by continuous variation impossible, leaving only discontinuous variation of the lines as a possible agent of evolutionary change.

Two features of Johannsen's work are worthy of note for us now. First, Johannsen himself was always deeply respectful of the work of the biometrical school. He never ceased to hold Galton in high esteem—he wrote in a popular presentation of his views in 1903 that "it now turns out that my material can be organized in such a way [taking natural populations as mixtures of pure lines] that it totally confirms Galton's law [of ancestral heredity]" (Johannsen, 1903). He endeavored to extend the same respect to Pearson. In March of 1905, Johannsen wrote to him that, "being eager to understand more clearly your methods in general and especially your calculation of Homotyposis, I should be glad to visit for some days or a week your laboratory as a student…not as a biological professor with proper meanings." He goes on to add that "I am sure that biologists ought to learn much more mathematics than now commonly [taught] to them, but also I am sure that mathematical visions ought to

FIG. 5.2

Wilhelm Johannsen, 1911.

Credit: Det Kgl. Bibliotek, Denmark (http://www5.kb.dk/images/billed/2010/okt/billeder/object461303/en/).

Public domain photograph taken in 1911.

take more care of the biological premises" (Johannsen, 1905). Pearson, for his part, would have none of it from someone who was willing to show even the slightest inclination toward Mendelian conclusions. He forwarded the letter on to Weldon, calling it "amusing." "I also said," Pearson writes (with his usual level of warmth), "that while I should be pleased to see him, I could not hope to teach him anything in the first four days of term, or in any four days at any time. I ventured also to say that I had no prejudice against 'pure lines' as such, but only against defective methods of demonstrating their existence. I wish he would stick to Bateson & leave me alone!" (Pearson, 1905, fol. 1r). So much for a collaboration between Johannsen and the biometricians.

Much as we saw above with Castle, however, this disagreement says nothing against Sturtevant's apt observation that Johannsen did much to advance the use of statistical methodology in biology. How, then, did he do so? Let's turn to the book which Sturtevant mentioned, the expanded German edition of Johannsen's *Elements of an Exact Theory of Heredity*, published in 1909, the same year as Bateson's textbook. From the start, Johannsen takes himself to have an explicitly methodological orientation, and places statistics at the center of those methods:

> The plan of these lectures is to give an elementary but critical presentation of the exact experimental theory of heredity as it has now been developed. [...] This is only possible, however, if the methods *that have been—or at least should have been—followed in the research are given special consideration. An essential aspect of these methods is their mathematical character, and can be described as applied mathematics. Familiarity with these methods is absolutely necessary for a real understanding of many hereditary questions.*[b]

(Johannsen, 1909, p. 1)

For Johannsen, then, as early as 1909 it is already the case that mathematical, and particularly statistical, training is essential in order to understand heredity. Of course, we must be careful to ground our mathematical work in the biological facts—as Johannsen himself puts it, "we must apply hereditary theory with mathematics, not as mathematics" (1909, p. 2)—but only when we can properly structure our data will we be able to see our way past the morass of contradictory observations in which our understanding of heredity usually finds itself.

As might be expected by his enthusiasm for the work of the biometricians, Johannsen was extremely well read in the applications of statistics to biology; in considering the utility of "Quetelet's law," he writes that "first and foremost we can mention Galton, and also the zoologists Weldon, Heincke, Duncker, Davenport, and others, the botanist Hugo de Vries, Ludwig, Verschaffelt, and many others, as well as the mathematician Karl Pearson, who worked out mathematical methods to test the agreement of a given series of variations with the binomial distribution" (Johannsen, 1909, p. 9). But the caution that he urges in the application of statistics to biological

---

[b] Translations from the German are my own, with grateful thanks to Thony Christie for comments and corrections.

theory is a recurrent theme, and he clearly believes that physiological and experimental work is conceptually prior to statistical analysis. While Galton was "the first to try to find certain laws" (Johannsen, 1909, p. 103) governing the relationship between the deviation of parents and the deviation of offspring, the biometricians' belief that "through continued selection one can gradually and substantially change the 'type' of the population concerned" (Johannsen, 1909, p. 112) clashed with Johannsen's work on pure lines, and meant that their efforts to connect heredity with evolution stood in dire need of critique. It is a fact, he writes, that *selection produces nothing*—rather, it "exterminates, creates space; this may be very important in natural life, but it does not interest the more narrowly limited research of heredity" (Johannsen, 1909, p. 463). As a result, he notes as he concludes the work that

> Adaptation *is a highly important physiological* fact, *inherent to the nature of the living organism, one might say. Adaptation, however, has not been shown to be "hereditary," i.e., the adaptations of the individual body do not noticeably influence the genotypic basis of the gametes of the individual concerned. In any case, [above] we did not find any evidence of "hereditary" adaptation.*
>
> **(Johannsen, 1909, pp. 463–464)**

Johannsen, then, like Castle, stands as a staunch critic of the theoretical portion of the biometrical outlook, while remaining firmly committed to the utility of statistical analysis for the interpretation of his data on pure lines.

In short, whatever the opinion of these authors about the importance of statistical tools for the interpretation of population change and natural selection, for an increasing number of experimental researchers, the methods of statistics were becoming irreplaceable by as early as 1910. Calls abound throughout these publications for more mathematical and statistical training for biologists, more advanced methods, and a tighter interplay between the use and development of applied work in mathematics and the practical work of breeding and biological experiment. Put differently, the end of the biometrical school by no means entailed the end of—or even the stagnation of—mathematical or statistical methodology in the biological sciences.

A mathematical theory of inheritance without statistics: The American school

In the absence of broad statistical training, however, a number of biologists during this period attempted to derive mathematized theories of inheritance that didn't rely on anything like the kind of statistical foundations that we would now recognize. A variety of authors thus produced theoretical efforts to understand both heredity and natural selection, but in such a way that the formalisms which resulted were only generalizable with difficulty, and which further made it extremely challenging to detect any theoretical commonality across different populations.

While there is a host of such work, I will again focus on two particular examples. First, let's consider the review article written by the aptly named Herbert Spencer

FIG. 5.3

Herbert Spencer Jennings, 1906.

Credit: Hinsdale, B.A., Demmon, I. (Ed.) 1906. History of the University of Michigan. p. 353. Public domain book, available from the Internet Archive at https://archive.org/details/historyuniversi00hinsgoog.

Jennings (Fig. 5.3) in 1916, summarizing several years of the development of the formalization of inheritance (Jennings, 1912, 1914; Wentworth and Remick, 1916; the effort would be continued by Robbins, 1917, 1918a, b, c) produced predominantly in America through the middle of the 1910s. Jennings describes the task at hand roughly as follows. Suppose that we start with two parents, one each homozygous for the dominant and the recessive character of a Mendelian trait. Suppose that they breed, and afterward, particular kinds of breeding regimes are instituted among their progeny. For instance, we might envision completely random mating, or completely assortative mating (in which individuals are only mated with other individuals of the same phenotype), or we could discard the recessive-phenotype individuals in every generation and breed from the remainder, and so on, and so on. After a finite number of generations, what would we expect the ratios of both the phenotypic characters and of the alleles within those populations to be? What would those populations look like in the infinite limit of long-term breeding, and how quickly would that limit be approached?

These kinds of results, as these authors quickly realized, can be ground out longhand using systems of recurrence equations—algebraic equations which define the constitution of the next generation in terms of the constitution of the current generation. Of course, we will still only get reasonably accurate predictions of real-world breeding results from such equations "if the experiment is carried out on an extensive scale, with many lines of descent" (Jennings, 1916, p. 55). But the level of sophistication that was obtained here was impressive, extending to such arcane systems of breeding as the analysis of long-run inbreeding experiments using sex-linked characters. This is all the more remarkable given that Jennings describes the procedure for deriving such formulae as one of painful guesswork and case-by-case verification. He writes that

> *One must, as a rule, take the series of results as finished products, and make an independent study of them, endeavoring by processes of trial to fit them to some series or to some formula. It is here that there is scope for ingenuity; a given series of results may resist for weeks the discovery of the law that unites them. [...] After a law or regular series is obtained that fits the first five or six generations, the law is applied to give the results for three or four generations more. These results are then tested by the actual detailed working out (symbolic formation of gametes and their mating, etc.)....*

(Jennings, 1916, p. 62)

It took a series of articles in 1917 and 1918 from Rainard B. Robbins (1918a, b, c, 1917), a mathematician at the University of Michigan who would later go on to a career as a recognized actuarial expert and vice president of the Teachers Insurance and Annuity Association (now better known as TIAA), to systematically apply methods of mathematical induction to the derivation of these sorts of formulas, and hence to prove that they were indeed correct in the long-run limit. The dozens of formulas which this method produced gave us useful rules of thumb for the construction of various kinds of agricultural breeding experiments, but no suitable basis from which to derive a general theory of evolution by natural selection.

The same sort of approach was applied in the case of the distribution of characters within populations in an unusual paper by Howard C. Warren (1917), an illustrious experimental psychologist who founded the Department of Psychology at Princeton. As he wrote, while "there appears to be no *inherent* incompatibility between Darwin's fundamental principles and the newer concepts of pure lines, mutations, and unit characters," we have nonetheless lacked appropriate methods for the testing of these claims. In particular, Warren argues that, insofar as the relevant data for the real examination of the efficacy of natural selection must be "the changes from generation to generation in the relative proportion of the individuals bearing alternative characters, in a representative population living under natural conditions," we need not only to collect that data in the wild, but further, to determine what we would expect to see if natural selection were in action; that is, we need to know "the effects of natural selection as obtained by calculation" (Warren, 1917, p. 306). To that end, Warren aims to compute the expected effect of natural selection under very narrow circumstances, namely, with complete Mendelian dominance, fixed fertility, a population at the environmental carrying capacity (hence, fixed population size), and natural selection represented by the dominant being twice as viable as the recessive or vice versa (hence, two values of fitness, expressed solely in terms of viability). Under these conditions, Warren finds that natural selection is indeed effective and closes his paper by calling for the experimental search for real-world circumstances that might instantiate the theoretical prediction that he has there derived.

Where Jennings and his colleagues had been aiming to derive formalisms for inheritance itself—and, when they considered selection at all, considered programs of artificial selection that would be operated by a human agent—Warren moves decidedly in the direction of a theory of evolutionary change itself, considering a selective influence that only partially favors one phenotype over the other. That said, his work shares with that of Jennings the feature of being painstakingly solved generation-by-generation—in Warren's case, only for three generations' worth of results.

Here, then, we see the continuation of another piece of the biometrical program, the desire to produce formalized theories of heredity and evolution. But because all of these authors were so focused on predictive results about the progeny of individual breeding pairs, and because the methods they used required such complex calculation, they found themselves unable to produce anything more than compendiums of short-hand formulas for the prediction of the results of various breeding regimens. It was, however, clear to all these biologists that for both practical reasons and theoretical ones (think of Warren's desire to offer the mathematical predictions necessary to test whether natural selection was acting in real-world Mendelian populations), the derivation of formalized theories of inheritance and selection would be vitally important to our understanding of evolution.

The speed of selection: R. C. Punnett

One further, and particularly odd, approach to natural selection deserves notice here. Reginald Crundall Punnett, who was Bateson's assistant, acerbic colleague in the fight against Pearson and Weldon, and whose name is now known to secondary-school biology students everywhere, wrote in 1915 a detailed book cataloging the phenomena of mimicry in butterflies. Mimicry had often been taken to be a phenomenon inexplicable as the result only of the selection of small, continuous variations—for how could an organism profit from looking only 5% like a poisonous rival?—but Punnett here spends effectively the entire book simply describing the nature of mimicry across dozens of species of butterflies (the book closes with 16 full-color plates).

The first appendix to this book, however, which goes entirely unreferenced in the main text, is something quite different (Punnett, 1915, pp. 154–156). In it, a local Cambridge mathematician, Harry Norton, derives a table containing a method for estimating the speed of population change brought about by natural selection. If we begin with a population consisting of particular percentages of an old variety, of hybrids, and of a new variety (11 such example population compositions are considered), establish whether the old variety or the new is dominant, and choose a value for the strength of selection (in this case, the chance of survival of the old variety and the chance of survival of the new variety; four possible strengths are included), the table tells us how many generations it would take to pass from one given fraction of the new variety within the population to another, assuming random mating. To reproduce one of Punnett's examples, if we start with a population in which the new variety is recessive, forms 2.8% of the population, and has a 10% selective advantage (i.e., 100% of the recessives survive, while only 90% of the dominants survive), then after 52 generations, the proportion of recessives will rise to one quarter.

The strangeness of this derivation and its presence in Punnett's book is difficult to overstate. The data it contains are presented quite poorly, and the description which Punnett provides is so convoluted that it doesn't offer much illumination, either. He makes no reference to the table anywhere else in the book, or does he describe in the appendix why he has included it, and it's not clear that he used the values it contained for any practical purpose. He argues in the main text that "the facts, so far as we at present know them, tell definitely against the views generally held as to the part played by natural selection in the process of evolution" (Punnett, 1915, p. 153), so it's further unclear why he felt that explicating the rate of natural selection was a particularly worthwhile enterprise, unless he had hoped that he would find definitive evidence of changes in populations of butterflies which were occurring faster than would be possible according to the values in the table. If he did, he does not offer them to us here.[c]

Historical curiosity though it may be, it is striking that even as staunch a Mendelian, Batesonian, and anti-selectionist as Punnett was in 1915 would have gone to the trouble to publish a formalization of the potential effects of selection. Again, we find yet more evidence that mathematical understandings of evolutionary theory were on their way to becoming quite commonplace.

Statistical inheritance in populations without selection

A general trouble, then, running throughout all of the treatments of heredity and evolution in the last two sections, is the unsuitability of their mathematical tools to the problems that they are intending to solve (a point that was also made for the case of Pearson by Morrison, 2002). That is, with the benefit of hindsight, we can now see that there is simply no generalization to be had about the behavior of evolving populations if one starts with the kinds of algebraic, recurrence-based approaches for particular populations that were proposed by these authors. That said, it is not as though statistical methods were only utilized for the analysis of experimental data during this period. Yet another group of theoretical workers were considering the application of statistical analysis to the theory of the transmission of continuously valued traits.

As had already been frequently noted as a theoretical possibility in the first few years after 1900, if a phenotypic character were, in fact, governed not merely by one, but by a very large number of Mendelian factors, then the corresponding number of available combinations of those Mendelian factors could produce the appearance of a character that had a continuous value. The first to develop this theory in any detail was the Swedish biologist Herman Nilsson-Ehle, who published an article noting that some non-Mendelian, apparently continuous inheritance phenomena in oats and

---

[c] Notably, J. B. S. Haldane would be inspired by this table in his later work of the 1920s, though consideration of Haldane falls unfortunately outside my scope here.

wheat could be thus analyzed (Nilsson-Ehle, 1908). Unfortunately, Nilsson-Ehle's paper was published in German in a poorly distributed Swedish botanical journal, and thus his insight was sporadically recognized and frequently "rediscovered."

One of the primary re-discoverers was the American biologist Edward Murray East (Fig. 5.4), who worked at Harvard alongside William Castle. In an article in *American Naturalist* and later an agricultural bulletin prepared with his doctoral student Rollins Emerson (an agriculturalist from the University of Nebraska), East, who learned about Nilsson-Ehle's paper only after he had begun to prepare the analysis of his own data, both re-derived the extant results regarding continuously distributed characters and confirmed them in maize (East, 1910; Emerson and East, 1913). In the latter work, Emerson and East introduce some representative examples of four-factor character traits at work in their maize plants, and then turn to their theoretical interpretation, pointing out the "two new and important phases of Mendelian inheritance" (1913, p. 13) that one could now consider given the possibility of highly multi-factorial Mendelian traits. The first is that such interactions of multiple genes

FIG. 5.4

Edward Murray East, 1922.

could explain "the possibility of having new characters formed under operation of the Mendelian law" (1913, pp. 13–14). While no such cases could yet be confirmed in nature, if the duplication of genes could yield cumulative effects (for instance, adding rows of kernels to an ear of corn), this might produce new degrees of quantitative or even new qualitative characters.

Further, and more importantly for our purposes, they directly connect the possibility of highly multi-factorial Mendelian inheritance to the production of statistically distributed or blending inheritance:

> …*when one considers the difficulty of distinguishing the zygotes having various [genotypic] formulae even when dominance is comparatively perfect, he might expect a population of [offspring] individuals with almost continuous quantitative variation if dominance is imperfect or absent. This gives a clue to a Mendelian interpretation of the inheritance hitherto known as blended.*
>
> **(Emerson and East, 1913, p. 14)**

They note as well that hypotheses about the number of genes responsible for a given, apparently continuous, character can be examined by investigating the variance of the character's distribution over time—apparent blending inheritance arising from many Mendelian factors will tend to increase its variance when inbred across generations, while the variance of "true" blending inheritance will remain the same, in the absence of any causal influence which might change it. They then present extensive data from maize indicating that a number of characters, including number of rows on an ear, ear size, plant size, and more, are in fact such Mendelian-segregating, continuous characters. "In general," they summarize, "it may be said that the results secured in the experiments with maize were what might well be expected if quantitative differences were due to numerous factors inherited in a strictly Mendelian manner" (Emerson and East, 1913, p. 107).

East's work thus combines a number of the features of biometrical evolutionary theorizing which we have had in view. Statistical understanding of inheritance, including careful analysis of the distributions of trait values across populations, is applied to our understanding of heredity across generations, in a way that's entirely consistent with (indeed, grounded in) Mendelian transmission of the underlying genes. On the other hand, and perhaps as a result of East's agricultural orientation, there is effectively no reference to natural selection in this work. He also seems to have no interest in presenting his results as a theory with implications for anything beyond this particular sort of continuously-valued characters in maize. The prevalence of blending inheritance, and the difficulty with which it was explicable on the kinds of simplistic Mendelian pictures available in in the first years immediately following 1900, were two significant reasons that the biometrical school continued to believe that Mendelism could not be valid. East seems neither to be interested in the ways in which his approach to blended inheritance could reduce the apparent distance between the biometrical and Mendelian schools, nor in the ways in which it might be applied to understand the population-level behavior of "traditional" Mendelian characters.

The exception to the rule: George Udny Yule

Again, let's take a step back and look at how we might summarize what we've seen so far. A number of researchers, with different presuppositions, different guiding influences in the literature, different training, and different goals, have all innovated on and continued to develop various pieces and combinations of the elements that defined the theoretical outlook of the biometrical school. One particularly important figure in this reconciliation with the Mendelian program has, however, yet to appear in our narrative: George Udny Yule (Fig. 5.5).

After he finished his degree and spent a year in training in physics in Bonn, the young mathematician Yule was offered a position as "demonstrator" for Karl Pearson in 1893—a sort of combined teaching and research assistant. He held this position for 3 years, until being promoted to a proper Assistant Professorship in 1896. After a brief stint (for financial reasons) outside the academy, he returned as Lecturer in Statistics at University College from 1902 to 1909, and finally as University Lecturer in Statistics at Cambridge from 1912 until 1931, when he resigned as a result of persistent medical issues (Yates, 1952).

FIG. 5.5

George Udny Yule.

Credit: Yates, F. 1952. George Udny Yule. 1871–1951. Obit. Not. Fellows R. Soc. 8 (21), 308–323, p. 308.

From our perspective, Yule is a peculiar figure, and I should begin by saying a bit about his role in the wider historiography. He was primarily a statistician by trade—his most lasting contribution in general is his much-reprinted textbook, *Introduction to the Theory of Statistics*, and the bulk of his papers are broadly technical. He is commonly recognized to have a significant role in the history of biology, but as a sort of standard footnote: Yule is taken to represent a kind of "aborted synthesis," an effort toward the recognition that Mendelism and biometry could be unified, prior to Fisher and other work in the early Synthesis, which fell, for various reasons, on deaf ears within the larger community of life scientists. Many proposals try to explain this neglect, and two are worth noting here. First, it's said that the acrimonious state of the debate itself simply rendered the main players unable to see the importance of Yule's work. Provine, for instance, writes that while Yule's "was the approach of population genetics," it was nonetheless the case that "conflicts among his contemporaries prevented its development at this time" (Provine, 1971, p. 89). Both sides could have—again the sense of regret—resolved their differences far sooner, had they only been able to follow Yule's example. Blinded by their mutual hatred, Pearson and Bateson's squabble sets back the history of science by several decades. Second, and not offering a much more illuminating account, it is sometimes claimed that Weldon and Pearson refused to consider Yule as a biologically adept peer, thus writing off his evolutionary work in advance. Much of the documentary burden here seems to rest on a single quote drawn from a letter from Weldon to Pearson from New Year's Eve, 1903. Weldon is commiserating with Pearson over an argument with Yule on the value of Johannsen's pure line experiments, discussed above. Weldon writes that Yule's "tone in controversy is never good, and although you say he is a good mathematician, it has never seemed to me that he grasped the elements of any biological problem" (Weldon, 1903, fol. 1r). Dismissive though this is, it stands alone in Pearson and Weldon's correspondence and is far outweighed by the apparent esteem in which they held the rest of Yule's work. I find it likely that this is an aberration brought on in defense of a friend, rather than any reasoned, negative judgment of Yule's worth as a biologist.

For my purposes, then, it is time to reassess the position and character of Yule's work, particularly in light of the other examples in this chapter, which have presented a view of biology in the 1900s and 1910s that is much more integrated, collaborative, and continuous than a single-minded focus on the biometry-Mendelism debate would have led us to expect. Yule's primary contribution to the field consists of an article published in two parts in 1902. Notably, Yule chose to print his article in a "neutral" journal, the very first volume of *The New Phytologist*, rather than taking it to the more tightly patrolled environs of either Pearson's *Biometrika* or Bateson's journals of the Royal Society. As we now are well aware, neither man was much interested in conciliation, and the young journal was therefore safer ground.

Yule's first goal is to clarify the stakes in the argument, and to correct what he saw as persistent misrepresentations by Bateson. He emphasizes, as I have above, that the biometrical perspective on evolution is fundamentally population-based, and not concerned with the lineages that arise from the matings of particular individuals. Thus, to indict the biometricians for having failed to consider the outcomes that would arise from a breeding experiment tracking the progeny of an individual instance of hybridization is simply to misunderstand what kinds of natural phenomena they were hoping to describe. Further, the biometrical formalism isn't applicable to just any group of organisms. To take one of Yule's examples, in a population that consisted of a mixture of a tall race and a short race, we could imagine a case where "there should be no individual heredity for either race taken separately"—that is, in which the individual organisms would not produce offspring differing from the mean of the tall or short type to which they belonged—"and yet the mixture would exhibit an apparent heredity [that is, variability around *its* mean] due simply to the constancy of type for each race" (Yule, 1902a, pp. 198–199). Galton and Pearson, Yule ventures to claim, simply would not (or, perhaps better, should not) have considered phenomena like the results of Mendelian crossing experiments to fall under the remit of their theoretical system. Before applying the tools of biometry, we must be sure that we have the right kind of population in hand.

Given this, we can start to see a way in which we could combine a Mendelian approach to the germ cells with the ideas of regression or reversion that underlie the law of ancestral heredity. "The ancestry of an individual," he writes, "may serve as guides to the most probable character of his offspring simply because they serve as indices to the character of his germplasm as distinct from his somatic characters" (Yule, 1902a, p. 206). The goal, then, if Mendel's laws and the law of ancestral heredity are theoretically compatible, "is to delimit their respective spheres, and shew in what way the one type of law may pass into the other, or the two even coexist" (Yule, 1902a, p. 207). He goes on to provide a demonstration of exactly this, showing that for a very particular set of mathematical constants in the law of ancestral heredity, we can recover exactly the ratio of dominant characters as derived in Mendelism (Yule, 1902b, pp. 226–227). For recessive traits, on the other hand, the result does not hold. But the fact remains that biometrical analysis can, at least in certain situations, can be applied to Mendelian phenomena. Here lies a sliver of the "special case" approach to Mendelism that had seemed so appealing to Weldon.

The general upshot that he hopes both sides will draw from his work is made quite explicit. First, Mendelism is important precisely because it demonstrates a biologically plausible mechanism of the transmission of characters:

> *The value of the work of Mendel and his successors lies not in discovering a phenomenon inconsistent with that law [of ancestral heredity], but in shewing that a process, consistent with it, though neither suggested nor postulated by it, might actually occur.*
>
> **(Yule, 1902b, p. 227)**

Biometry, in contrast, needs to seek further flexibility, so as to derive formalisms that can successfully encompass more, and more peculiar, Mendelian phenomena.

> *What is required from a physical theory of heredity is that it should assign a meaning to the variations in the constants that do occur, enabling one, given the law of ancestral heredity for an organ, to state the relative influences thereon of the different agencies concerned—selection, in all forms, circumstance, and so forth.*
>
> **(Yule, 1902b, p. 237)**

In this sense, then, the standard story about Yule is exactly right: he is clearly anticipating a sort of synthesis here, a way in which some further flexibility with mathematical formalism on the part of the biometricians, combined with similar methodological flexibility on the part of the Mendelians, could result in a generalized theory of heredity. No such theory, he readily admits, is available in 1902—but he finishes his work on a hopeful note, claiming (quite rightly, in hindsight) that there is little reason that we should not expect to be able to derive one quite soon.

It is interesting that Pearson and Bateson both immediately and entirely rejected these conclusions. Bateson's jibe at those (like Yule) seeking "some convenient shelter of neutrality" by arguing for the compatibility of Mendelism and biometry has already been noted above, and Pearson went so far as to engage in a few articles worth of published debate over Yule's results (Pearson, 1907b, 1904b; Pearson et al., 1903; Yule, 1906). Yule, after a few further arguments with Pearson over correlation tables, essentially leaves the study of heredity entirely, spending the rest of his career writing in pure statistics, his only later biological work concerning results in ecology about laws of distribution of species and the nature of species ranges (somewhat similar to Darwin's work through the middle chapters of the *Origin*).

Other than (once again) simple personal antagonism, what reasons might Bateson and Pearson have had for rejecting Yule's proposal? As has been compellingly argued by James Tabery (2004), this "rejection" narrative needs to be carefully nuanced. The beginning of Pearson's immediate reply to Yule's work in early 1903, a hastily added appendix to a paper already in press for *Biometrika*, starts by noting that he agrees with much of Yule's work, "for example, with his insistence on the point that the laws of intra-racial heredity are not incompatible with Mendelian principles holding for hybridisation" (Pearson et al., 1903, p. 228). But he disagreed with the further conclusions which Yule drew, and Tabery argues that both Pearson and Bateson had their own clear and compelling reasons for disputing Yule's work. For Pearson, the extra flexibility that Yule demanded in the law of ancestral heredity simply ran counter to his intuition that one form of that law, with a single, fixed set of numerical constants, should hold across all evolving populations. Pearson, that is, had no interest in being flexible in the way that Yule's approach required. Bateson's skepticism, according to Tabery, is more complex. On the one hand, Yule clearly argued that the law of ancestral heredity was the fundamental phenomenon, and the Mendelian forms of inheritance which Bateson prized were a special case, a direction of explanatory dependence that would not have sat well with Bateson. Second, Tabery notes that Yule's distinction between two types of heredity—that within a "race" and that between "races"—holds only if we have a clear and consistent definition of the term "race," which Yule himself never provides. Finally, Tabery suggests, in a very similar critique to one offered by Castle a century earlier (Castle, 1903, p. 234), that Yule misunderstands the nature of Mendelian inheritance itself.

Castle and Tabery have both misread Yule here, however, so it is important to spend a moment on Yule's formulation of the compatibility of Mendelian inheritance and biometry. As described above, Yule is considering the ways in which the question of individual inheritance, as treated by the Mendelians, might be related to the questions of intra-racial heredity and population inheritance that concerned the biometricians. He asks us to examine a population composed of half homozygous dominant and half homozygous recessive organisms. He notes that in one generation they will reach (the not-yet-coined) Hardy–Weinberg equilibrium, and afterwards the trait distribution in the population will remain static. What would then happen, he wonders, if we were to consider just the individuals that showed the dominant phenotype within that population to be a separate "race," and analyze the dynamics of their "population" change under random mating using the methods of the biometricians?[d] Imagine that there were 400 individuals in the population as a whole; by classic Mendelian ratios 300 of them would exhibit the dominant phenotype, 100 pure-dominant and 200 hybrid. As the population continued to mate randomly in the future, what would *their progeny* look like over time? Yule notes that the 100 pure dominants would give rise to 50 pure dominants and 50 hybrids, and the 200 hybrids would give rise to 50 pure dominants, 100 hybrid dominants, and 50 recessives. All standard Mendelian fare. On the whole, then, this means that the first-generation offspring of the dominants would be five-sixths dominant, one-sixth recessive. It is this very particular scheme, extended over multiple generations, in which Yule proves his result: the dynamics of the dominant phenotype within a randomly interbreeding population initially formed of hybrids will instantiate one form of the law of ancestral heredity.

Yule is thus engaging in precisely the kind of work that one would expect from an effort to derive a general-purpose theory of evolution which exemplified the four desiderata that I laid out at the beginning of the chapter. He essentially takes the occurrence of Mendelian transmission as writ, at least in some (now experimentally well-confirmed) circumstances, and is endeavoring to show a precise way in which a particular population composition could have led to the appearance of biometrical laws in a case which remains entirely compatible with that underlying Mendelian inheritance pattern. While he cannot yet offer a theory of natural selection on this basis, he is confident that we soon will be able to do so.

The view from the textbooks

What we have here, then, are a number of threads in the literature, ranging from the adoption of only a few of the tenets of the biometrical outlook to the near complete anticipation of the synthesis of biometry with Mendelism to which we will finally turn in the next chapter. There is a pervasive sense that the possibility, and even the desirability, of such a synthesis was much more "in the air" than would be acknowledged by the traditional histories of this period, and that such a synthesis owed more to and exhibited more continuity with the works of the biometricians than is commonly held to be the case.

---

[d] This is where Castle and Tabery's reading of Yule goes wrong: Yule is asking us to consider the dynamics of the progeny of dominant-phenotype individuals within the *larger* population (i.e., mating randomly in the larger population), *not* to only consider the breeding dynamics of an isolated population composed of dominant individuals.

That said, one important and powerful objection remains. As I have presented them here, these are scattered breadcrumbs, finding their expression in highly divergent literatures in experimental biology, genetics, mathematics, and even psychology, across a period of time measuring around a dozen years and on multiple continents. Surely, one might argue, no young biologist in training would have simultaneously had access to, much less been able to digest, all of these sources. I have, the objector might claim, forced an image of continuity and commonality where one could not possibly have existed for actors on the ground.

Of course, it is impossible to conclusively refute such a convinced objector, except possibly by extremely detailed study of the course syllabuses and individual reading habits of particular biologists who were trained during this period. But one further piece of evidence should be telling in this regard: the textbooks on inheritance and evolution that were produced during this period (we will consider examples from 1906 to 1912) should offer us a window into the ways in which general treatments of evolution and heredity presented the state of play. I want to close this chapter, then, by considering three such textbooks; all, we will find, are entirely in accord with the broad story that I have outlined here.

Robert Heath Lock's *Recent Progress in the Study of Variation, Heredity, and Evolution* (1906)

Robert Heath Lock (Fig. 5.6) graduated with a degree in botany from Cambridge in 1902, having taken lectures from Bateson himself. Newly appointed a Cambridge fellow alongside his colleague Punnett, he set himself about writing a textbook covering evolution, heredity, and genetics, which was published in 1906 and later released in four further editions.[e] He made his career as a director of botanical gardens the world over, starting with the Cambridge University Herbarium and later including time in Sri Lanka and Rio de Janeiro. He died prematurely of influenza in 1915, at the age of 36, leaving behind his wife, the writer Bella Woolf (sister-in-law to the more famous author, Virginia).

The scope of Lock's book is impressive: he covers organic and inorganic evolution, natural selection, biometry, mutation theory, the study of hybridization before Mendel, Mendelism, cytology, and contemporary work on continuous variation, pure lines, and human evolution. For our purposes, though, what is interesting is his nuanced approach to the relationship between biometry and Mendelism.

Lock on multiple occasions presents a classic, Galtonian picture of the derivation of the laws of chance. Coin tosses, he writes, are a useful tool for considering such laws precisely because they illustrate "the more general assumption of an event or magnitude depending upon a number of causes of equal strength, which in the long-run act with equal frequency in two opposite directions" (Lock, 1906, p. 88). Given that this is sufficient to license the use of the methods of statistics, we should

---

[e] Lock is a little-studied figure in the history of genetics; most of the biographical information here follows Edwards (2013).

FIG. 5.6

Robert Heath Lock.

Credit: Lock, R.H. 1920. Recent Progress in the Study of Variation, Heredity, and Evolution, sixth ed., p. xii.
Public domain image.

expect them to be of great utility in biology—at the very least, for the understanding of *acquired* variations. "During the development of an individual a great number of different external influences come into play, leading to slight modifications of every part, now in one direction, now in another. This being so, we may be quite sure that a large proportion of the normal variability which any species exhibits is acquired" (Lock, 1906, p. 283).

For variation in the germ cells, however, things are more difficult. On the one hand, the use of statistics observationally will clearly remain important. If all we mean in statistically describing a population is to claim that "a more or less definite numerical value can be attached to the average amount of resemblance between any specified pair of relatives" (Lock, 1906, p. 103), then we can derive a number of meaningful results, such as the constancy of the rate of inheritance (i.e., the similarity between parents and offspring, as Galton had studied) in man and in animals, as well as the apparent similarity of inheritance of mental and physical characteristics. On the other hand, if we want to extend our consideration to the more complex results of the biometrical school, including the ascription of normal variation to properties of the germ cells themselves, Lock notes that these are often "based upon the assumption that the law of ancestral heredity is strictly true," and hence different in character from "the facts of normal variability and of correlation between relatives" (Lock, 1906, p. 106).

This means that we will have to work harder to bring them into accord with other facets of contemporary biology.

Lock considers two promising ways of doing so. First, he extensively discusses the work of Yule. While, he writes, the biometrical school (that is, Pearson) has yet to pass final judgment on whether or not it considers those derivations to be valid, "Mr. G. Udney [sic] Yule endeavoured with some apparent success to reconcile the Mendelian results with those of biometry. Progress has been rapid during the last four years, and what we now have before us is rather the question of reconciling the biometrical conclusions with the firmly-established facts of Mendelian inheritance" (Lock, 1906, pp. 209–210). Yule's type of analysis, Lock claims, is certainly one way in which we might hope to do so.

Second, Lock discusses Johannsen's results, particularly his derivation of biometrical results from a population consisting of a mixture of pure lines.

> *There is no reason to doubt that the statistical treatment of such a population would yield similar results to those actually obtained by biometricians from the data at their disposal.... [Such a population] would admirably fulfill the conditions we have seen to be necessary in the case of material to which methods based upon the theory of chance are to be applied. The phenomena which follow [upon crossing] still remain to be worked out, but it is not unlikely that they will be found to conform to those laws of heredity associated with the name of Mendel....*
>
> **(Lock, 1906, pp. 110–111)**

This would constitute yet another combination of a type of populations and theoretical resources, for which we might find that phenomena described by the law of ancestral heredity could be equally well encompassed by Mendelian analysis. In short, Lock seems to see no inherent incompatibility between a biometrical explanation and a Mendelian one, provided that we can find the proper manner of connecting the two in a given evolutionary case.

That said, he is cautious of the excessive faith in biometry as expressed, for example, by the early Weldon we saw in Chapter 3:

> *Some students of biometry, however, would go very much further than this, for it is their professed position that their own form of study is the only method by which any real advance in our understanding of the processes of evolution can be brought about. This opinion is based upon the assumption, of which proof is wanting, that new species have arisen exclusively through the accumulation by natural selection of variations of a strictly indefinite, fluctuating, or normal kind.*
>
> **(Lock, 1906, pp. 74–75)**

Whatever may be said about the relation between biometry and Mendelism, Lock argues, we certainly have enough experimental data concerning discontinuous modification to make this kind of conclusion, on which discontinuous variation has no role to play whatsoever, highly doubtful.

In the scheme of Lock's entire text, however, this is a fairly minor quibble. We find here very little evidence of conflict or open warfare, nor of the complete and comprehensive victory of the Mendelians—rather, we see a strong desire to search for particular population or inheritance structures the analysis of which might open up space for reconciling biometry with Mendelism, both of which, Lock claims, have proven themselves as tried and useful tools for the analysis of inheritance and evolution.

J. Arthur Thomson's *Heredity* (1908)

Sir J. Arthur Thomson (Fig. 5.7) is a rather strange figure in the history of biology. He studied at Jena under Ernst Haeckel, eventually returning to Edinburgh, where he spent his career as the Regius Chair of Natural History at Aberdeen (Ritchie, 1933). Perhaps as a result of his early German training, he held an affinity for a number of what we might call "recessive" approaches to the study of evolution during the period. Bowler has examined a number of his works in detail, laying out his fondness for a progressionist understanding of evolution (Bowler, 2001, p. 44); a teleological, "anti-mechanistic view of life," drawing heavily from Bergson, that could reconcile science with religion (Bowler, 2007, p. 148, 2001, p. 137); a "holism" which drew on non-physical (especially psychological) explanations for organismic behavior

FIG. 5.7

Sir John Arthur Thomson.

Credit: Jordan, D.S. 1922. The Days of a Man, vol. 2, p. 548. Public domain book, available from the Internet Archive at https://archive.org/details/daysofmanbeingme02jordrich.

(Bowler, 2001, p. 170); and the altruism of Kropotkin (Bowler, 1992, p. 56). Despite his obituarist's depiction of his "gift of simple, lucid writing" (Ritchie, 1933, p. 296), his textbook *Heredity* is both too long and too florid. It bears little, however, of the iconoclastic streak that Bowler finds in his more popular writings. On the contrary, it aims at near-complete coverage of the status of investigations of heredity as of 1908—Bowler describes it as a "highly regarded survey," noting that it appeared in a book series containing a number of other cutting-edge, specialist works (Bowler, 2009, p. 117). The amount of experimental data that Thomson compiled, arguing both for and against nearly every widely held theoretical interpretation of inheritance, is quite impressive. Bowler cites his treatment of Lamarckism, for instance, as the most systematic analysis then available of the evidence for and against the inheritance of acquired characters (Bowler, 1992, p. 61).

Thomson at the outset records his debt to the works of "Galton, Weismann, Pearson, Bateson, and De Vries" (Thomson, 1908, p. viii), along with his conviction that "it is unanimously agreed that a satisfactory study of inheritance demands a minute inquiry into the history of the germ cells, a statistical study of the characters of successive generations, a careful criticism of the older data and of popular impressions, and a testing of hypotheses by experimental breeding" (Thomson, 1908, p. 17). In short, cytology, biometry, and Mendelism stand on an equal footing (along with careful critique) as aspects essential to the success of any future evolutionary theory.

He begins his discussion of statistical approaches by introducing the prior work of the biometricians. The population-level curves of fluctuating variation which were the stock-in-trade of biometry, he writes, "especially if made year after year, may show the direction in which the species is moving, perhaps the way in which selection is working, perhaps even that the species is splitting up into two subspecies" (Thomson, 1908, p. 79). He goes on to quote extensively from the editorial introduction to the first volume of *Biometrika*. Much later on, he notes that "when we have to study results that depend upon numerous complicated conditions, the statistical method is of special service. Not that it can ever tell us how the conditions lead up to the results, but it will tell us what regularity there is in the occurrence of the results, and…it may put us on the track of discovering the mechanism that connects them" (Thomson, 1908, pp. 310–311). No skepticism of mathematical approaches to evolution is here to be found, then. In fact, a statistical approach to evolution will allow us to discover the regularities crucial for later coming to a knowledge of causal mechanisms—one is reminded of Weldon's insistence that statistical observation is prior to and fundamental for later comprehension of the mechanistic-causal details of water intake in his crabs.

Thomson argues, precisely as Lock did above, that a purely biometrical perspective, of the sort for which Weldon and Pearson were pushing in the 1890s, is an antiquated view, insofar as it too tightly connects the theory of evolution to the requirement that new species be generated by small, normally distributed, individual variation. "Now," on the contrary, "there is no need to hamper the Evolution Theory by restricting selection to minute variations. We know that

sports, mutations, or discontinuous variations are frequent, and that they are re-
markably stable in their hereditary transmission" (Thomson, 1908, p. 81). Again,
we must seek ways in which to bring the two theoretical perspectives together.
Developing a novel theoretical vocabulary for the depiction of heredity, then, "is
all the more justifiable since we cannot doubt that all the ordinary phenomena
are of a piece, that many of the ordinary modes will be embraced eventually in
one general formula—probably some modification of Galton's Law of Ancestral
Inheritance, and that others will be embraced in Mendelian formulae" (Thomson,
1908, p. 109).

For Thomson, like Lock, one such promising avenue to found a general theory of heredity is the work of Yule, which we need not recount yet another time.
Another, discussed in the last chapter, is recent work on chromosomes. "It is interesting to notice," he writes, "that whether we consider Weismann's theory of the
determinants composing the germ-plasm, or the Mendelian theory of the segregation of characters in the germ cells, or De Vries's Mutation Theory, we are led
to the theoretical conception of elementary units [of inheritance]. And again, we
find the late Professor Weldon referring to Galton's Law" to describe to the same
kind of fundamental units, with Thomson quoting from the anonymous accounts of
Weldon's final series of public lectures printed in *The Lancet* (Thomson, 1908, p.
91). Biometricians and Mendelians alike thus seem to grant that the fundamental,
underlying nature of heredity is to be found in transmissible unit characters. While
Thomson recognizes that there is as yet no way on this basis to derive a theory en-
compassing the insights of both groups, he is insistent that there is also no reason
to reject the law of ancestral heredity. "It appears to us," he writes, "that there are
not a few cases where Mendelian interpretations do not work, and where a theory
of ancestral contributions, more numerous and more conspicuous in proportion to
the nearness of the ancestors, is at present justifiable…. [T]here must be something
in individual heredity to account for it" (Thomson, 1908, p. 333). For Thomson as
for Lock, then, biometry and Mendelism are two bodies of established fact to be
reconciled, and recent work gives us every reason to think that such efforts will be
successful.

Edwin S. Goodrich's *The Evolution of Living Organisms* (1912)

The author of our third and final textbook, Edwin S. Goodrich (Fig. 5.8), switched
careers from art to zoology, under the influential tutelage of E. Ray Lankester. He
followed his advisor to Oxford as his assistant in 1891, and proceeded to publish several articles as an undergraduate and aid Lankester in the restructuring of the Oxford
University Museum's zoological collections. He became a consummate and heralded
comparative anatomist, elected to the Royal Society in 1905, and worked first under Weldon and then G. C. Bourne, who each held the Linacre Chair of Zoology at
Oxford. He finally took that chair himself in 1921, where he remained for the rest
of his career (De Beer, 1947; Hardy, 1946; for his later career, see Morrell, 1997).
The marine biologist Alister Hardy wrote in his obituary of Goodrich that "while his

FIG. 5.8

Edwin S. Goodrich.

    Credit: de Beer, G.R. 1946. Edwin Stephen Goodrich. Journal of Anatomy 80, 112–113, p. 112–1.

research was mainly concerned with tracing the course of evolution it would be the greatest mistake to suppose that he was not interested in the causes underlying the process" (Hardy, 1946, p. 347). Goodrich's book on evolutionary theory, first appearing in 1912 and later reprinted in 1920 and expanded in 1924, was written for an introductory or popular audience; Hardy calls it, even after the advent of much of the Modern Synthesis, "still one of the best introductions to evolution for the student" (Hardy, 1946, p. 347).

I can discuss this work more briefly, only because it looks quite similar to the last two texts already summarized. Biometry has demonstrated its utility, Goodrich writes, in the analysis of variations in populations. "These variations can often be accurately measured, and the statistical study of variation begun by Quetelet and Galton, and carried on by W. F. R. Weldon, K. Pearson, and others, has yielded many important results" (Goodrich, 1912, p. 29). Johannsen's bean experiments are taken by Goodrich to precisely echo the corresponding biometrical results, and merely express a different way of arriving at "the 'law of ancestral inheritance' worked out by Galton and modified by Pearson. Pearson has defined it as a rule for predicting the average value of a character in the offspring from the value of the character in the ancestors" (Goodrich, 1912, p. 62).

The next move that Goodrich makes, however, is very interesting. Quite separate from the fact that the law of ancestral heredity can be shown to hold in populations which are a mixture of numerous pure lines, the fact that this is the underlying nature of the population at issue has an important further consequence. Namely, "the contributions to inheritance, therefore, from distant ancestors are negligible, and selection through very few generations is sufficient to yield a practically pure and constant race" (Goodrich, 1912, p. 63). It is unclear what Pearson and Weldon's response to this claim would be. Weldon until his death argued against the theory of the pure gamete, whether in Johannsen's pure-line or Bateson's Mendelian form. It is thus unlikely that he would have been in full agreement with the idea that a biometrical population could be treated as a mixture of such pure lines if he did not believe nature contained any such thing. In any case, it seems as though neither Weldon nor Pearson spent much time considering the rate at which a population would reach stability under the steady influence of natural selection, this pre-occupation having become a feature of the biological literature in the period after Weldon's death. They might, then, be willing to grant the point that the approach to stability is relatively quick, if never complete—at least in some real-world cases.

However, one might resolve this technical point, though, one can once again read Goodrich's textbook and come away with no idea that there was ever a struggle between biometry and Mendelism, and certainly with no idea that Mendelism was thoroughly victorious.

Textbooks to syntheses

Having examined three important, introductory-level presentations of evolutionary theory and heredity, in addition to a number of the primary sources upon which they drew, a few claims seem entirely justifiable in summary. First, approaching the last years before the advent of the Modern Synthesis exclusively through the lens of a battle between biometry and Mendelism, won by the Mendelians around 1906, seems entirely indefensible. Both specialist and introductory discussions of the understanding of heredity at the time present biometry and Mendelism on roughly equal footing—as two bodies of extremely successful theoretical and practical knowledge which need to be integrated if we are to have any chance of developing a general theory of the process of evolution. A variety of experimental and theoretical work, then, is taken to be central to this process: Yule's theoretical derivation of Mendelian results from biometrical premises, Johannsen's work on pure lines, and the possibility of derivations of population behavior from the underlying transmission of unit characters in the chromosomes are all extensively discussed, and all taken to be promising avenues for future research.

Further, and equally important for the story that I want to tell here, all four of the desiderata of the late biometrical research program—a *statistical* theory of *natural selection* at the *population level*, consistent with *Mendelian transmission*—seem to be quite widely shared by theoreticians during this period. Of course, a number of works make an advance only on some one or two of these axes, but one gets the pervasive sense in reading a variety of publications across this time that all four were taken to be essential to the future success, in particular, of a Darwinian, natural-selection-based approach to evolutionary theory.[f] To make the point in well-known Kuhnian terms, biological theorizing in the first two decades of the twentieth century resembles not a pre-paradigmatic war of all against all, nor a period in which a non-selectionist, exclusively Mendelian paradigm dominates, but rather one in which a profoundly normal science is taking place, as actors attempt, each in their own ways, to figure out how to bring together the numerous, fruitful, and rapidly developing varieties of theory and techniques of experiment that have simultaneously arrived on the scene.

One element, however, of the biometrical school seems not to have infiltrated these decades. I was at pains throughout the last two chapters to argue that biologists like Weldon and Pearson not only spent time developing theory and experimental practice, but also in developing a philosophical approach that could ground the ways in which they thought about the introduction of chance and statistical methodology into evolutionary theory. They cultivated, in the end, sophisticated thinking about the reasons for which statistical theorizing was required, and what those reasons had to say about the composition and nature of the biological world. Much of that kind of insight has (unfortunately, from the perspective of contemporary history and philosophy of science) fallen away in this era, and thus it is much more difficult to see what might have grounded the relationship between chancy theories and the biological world. When we do see discussions of the underlying conceptual foundations for the use of chance in evolution, they tend to be Galtonian rather than Weldonian—that is, they make fairly facile reference to things like the law of errors or the law of large numbers, or they appeal to the derivation of the normal curve from causal processes that consist of a large number of equally balanced shocks in opposing directions (think of Galton's quincunx device from Chapter 2).

There are, I think, a number of good historical reasons for why this might be the case. As we have seen, most of Weldon and Pearson's more sophisticated philosophical thought about chance was unfortunately never published. When it was, it often was accompanied by mathematics that many of the biologists working in this period were still ill-equipped to follow—a trend that would begin to change as the frequent calls for more mathematical education in biology began to be heeded. This means that much of the conceptual architecture for the further development of chance in evolution owes itself less to the more interesting work of Pearson and Weldon than to the popular presentations of Galton in *Natural Inheritance*. I thus want to present a carefully nuanced account of the kind of continuity that I see between the biometrical school and the early days of the Modern Synthesis.

At the conceptual level—in matters of biological theorizing itself—there is an extremely high degree of continuity, and the care with which the authors we have seen in this chapter were attempting to combine biometrical with Mendelian insights is impressive. At the axiological level—in matters of the stated goals of biological

---

[f] I lack the space to consider the ways in which non-Darwinian modes of evolution such as neo-Lamarckism or orthogenesis, the popularity of which has been insisted upon most notably by Peter Bowler (1992), might offer us alternative stories here.

theorizing—there is also a high degree of continuity, as I have already stressed. At the philosophical level, on the other hand—in matters of the picture of the theory-world relationship which gives rise to those stated, practical goals for biological theorizing—we have less solid evidence of potential continuity.

As we will soon see, there is indeed a documentary basis for claiming that this proto-synthetic view of evolutionary theory was more than coincidentally connected to early work in the Modern Synthesis, but we will need to cautiously and critically explore the possibility of connections of this sort. If one stands convinced of the alternative picture of biology during the period between the advent of Mendelism and the construction of the Modern Synthesis that I have laid out here, we still must trace the ways in which the views of evolution developed between 1906 and 1918 influenced biology to come. It is finally time, then, that we turn toward the end of our story, and the development of the early works of the Modern Synthesis, where—and here we can agree with the traditional historical narrative—the probabilistic and statistical nature of evolutionary theory was permanently cemented.

References

Ankeny, R., 2000. Marvelling at the marvel: the supposed conversion of A.D. Darbishire to Mendelism. J. Hist. Biol. 33, 315–347. https://doi.org/10.1023/A:1004750216919.

Bateson, W., 1909. Mendel's Principles of Heredity. Cambridge University Press, Cambridge.

Bowler, P.J., 1992. The Eclipse of Darwinism: Anti-Darwinian Evolution Theories in the Decades Around 1900. Johns Hopkins University Press, Baltimore, MD.

Bowler, P.J., 2001. Reconciling Science and Religion: The Debate in Early-Twentieth-Century Britain. University of Chicago Press, Chicago.

Bowler, P.J., 2007. Monkey Trials and Gorilla Sermons: Evolution and Christianity from Darwin to Intelligent Design. Harvard University Press, Cambridge, MA.

Bowler, P.J., 2009. Science for All: The Popularization of Science in Early Twentieth-Century Britain. University of Chicago Press, Chicago.

Castle, W.E., 1903. The laws of heredity of Galton and Mendel, and some laws governing race improvement by selection. Proc. Am. Acad. Arts Sci. 39, 223–242. https://doi.org/10.2307/20021870.

Cock, A.G., 1983. William Bateson's rejection and eventual acceptance of chromosome theory. Ann. Sci. 40, 19–59. https://doi.org/10.1080/00033798300200111.

De Beer, G.R., 1947. Edwin Stephen Goodrich, 1868–1946. Obit. Not. Fellows R. Soc. 5, 477–490. https://doi.org/10.1098/rsbm.1947.0013.

East, E.M., 1910. A Mendelian interpretation of variation that is apparently continuous. Am. Nat. 44, 65–82.

Edwards, A.W.F., 2013. Robert Heath Lock and his textbook of genetics, 1906. Genetics 194, 529–537. https://doi.org/10.1534/genetics.113.151266.

Emerson, R.A., East, E.M., 1913. The Inheritance of Quantitative Characters in Maize, Bulletin of the Agricultural Experiment Station of Nebraska. University of Nebraska, Lincoln, Nebraska.

Gigerenzer, G., Swijtink, Z., Porter, T.M., Daston, L., Beatty, J.H., Krüger, L., 1989. The Empire of Chance: How Probability Changed Science and Everyday Life. Cambridge University Press, Cambridge.

Goodrich, E.S., 1912. The Evolution of Living Organisms, first ed. T.C. & E.C. Jack, London.

Hardy, A.C., 1946. Edwin Stephen Goodrich, 1868–1946. Q. J. Microsc. Sci. 87, 317–355.

Jennings, H.S., 1912. Production of pure homozygotic organisms from heterozygotes by self-fertilization. Am. Nat. 46, 487–491.

Jennings, H.S., 1914. Formulae for the results of inbreeding. Am. Nat. 48, 693–696.

Jennings, H.S., 1916. The numerical results of diverse systems of breeding. Genetics 1, 53–89.

Johannsen, W., 1903. Om Darwinismen, set fra Arvelighedslærens Standpunkt [About Darwinism, seen from the point of view of the science of heredity]. Tilskueren, 525–541. http://www.bshs.org.uk/bshs-translations/johannsen, BSHS Translations. (Accessed 22 September 2021).

Johannsen, W., 1905. Letter from WJ to KP, 1905–03.

Johannsen, W., 1909. Elemente der exakten Erblichkeitslehre. Gustav Fischer, Jena.

Lock, R.H., 1906. Recent Progress in the Study of Variation, Heredity, and Evolution. John Murray, London.

MacCurdy, H., Castle, W.E., 1907. Selection and Cross-Breeding in Relation to the Inheritance of Coat-Pigments and Coat-Patterns in Rats and Guinea-Pigs. Carnegie Institution, Washington, DC.

MacKenzie, D.A., 1981. Statistics in Britain, 1865–1930: The Social Construction of Scientific Knowledge. Edinburgh University Press, Edinburgh.

Morrell, J., 1997. Science at Oxford, 1914–1939. Oxford University Press, Oxford.

Morrison, M., 2002. Modelling populations: Pearson and Fisher on Mendelism and biometry. Br. J. Philos. Sci. 53, 39–68. https://doi.org/10.1093/bjps/53.1.39.

Nilsson-Ehle, H., 1908. Einige Ergebnisse von Kreuzungen bei Hafer und Weizen. Bot. Notiser, 257–294.

Pearson, K., 1904a. A Mendelian's view of the law of ancestral inheritance. Biometrika 3, 109–112. https://doi.org/10.2307/2331528.

Pearson, K., 1904b. Mathematical contributions to the theory of evolution. XII. On a generalized theory of alternative inheritance, with special reference to Mendel's laws. Philos. Trans. R. Soc. Lond. A 203, 53–86. https://doi.org/10.1098/rsta.1904.0015.

Pearson, K., 1905. Letter from KP to WFRW. 1905-04-10.

Pearson, K., 1907a. On heredity in mice from the records of the late W. F. R. Weldon. Part I. On the inheritance of the sex-ratio and of the size of litter. Biometrika 5, 436–449. https://doi.org/10.2307/2331690.

Pearson, K., 1907b. [Review of] on the theory of inheritance of quantitatively compound characters on the basis of Mendel's laws, by G. Udny Yule. Biometrika 5, 481–482. https://doi.org/10.2307/2331701.

Pearson, K., 1908. Letter from KP to FG. 1908-12-10.

Pearson, K., 1909a. The theory of ancestral contributions in heredity. Proc. R. Soc. Lond. B 81, 219–224. https://doi.org/10.1098/rspb.1909.0018.

Pearson, K., 1909b. On the ancestral gametic correlations of a Mendelian population mating at random. Proc. R. Soc. Lond. B 81, 225–229. https://doi.org/10.1098/rspb.1909.0019.

Pearson, K., 1910. Darwinism, biometry, and some recent biology. I. Biometrika 7, 368–385. https://doi.org/10.2307/2345390.

Pearson, K., 1911. Further remarks on the law of ancestral heredity. Biometrika 8, 239–243. https://doi.org/10.2307/2331450.

Pearson, K., 1930. On a new theory of progressive evolution. Ann. Eugenics 4, 1–40. https://doi.org/10.1111/j.1469-1809.1930.tb02072.x.

Pearson, K., Blanchard, N., Lee, A., Lee, A., 1903. The law of ancestral heredity. Biometrika 2, 211–236. https://doi.org/10.2307/2331683.

Provine, W.B., 1971. The Origins of Theoretical Population Genetics. Princeton University Press, Princeton, NJ.

Punnett, R.C., 1915. Mimicry in Butterflies. Cambridge University Press, Cambridge.

Ritchie, J., 1933. Sir J. Arthur Thomson. Nature 131, 296. https://doi.org/10.1038/131296a0.

Robbins, R.B., 1917. Some applications of mathematics to breeding problems. Genetics 2, 489–504.

Robbins, R.B., 1918a. Applications of mathematics to breeding problems II. Genetics 3, 73–92.

Robbins, R.B., 1918b. Some applications of mathematics to breeding problems III. Genetics, 375–389.

Robbins, R.B., 1918c. Random mating with the exception of sister by brother mating. Genetics 3, 390–396.

Stigler, S.M., 1986. The History of Statistics: The Measurement of Uncertainty before 1900. Cambridge, MA, Belknap.

Sturtevant, A.H., 2001. A History of Genetics, second ed. Cold Spring Harbor Laboratory Press, Cold Spring Harbor, NY.

Tabery, J.G., 2004. The "evolutionary synthesis" of George Udny Yule. J. Hist. Biol. 37, 73–101. https://doi.org/10.1023/B:HIST.0000020390.75208.ac.

Thomson, J.A., 1908. Heredity. John Murray, London.

Warren, H.C., 1917. Numerical effects of natural selection acting upon Mendelian characters. Genetics 2, 305–312.

Weldon, W.F.R., 1903. Letter from WFRW to KP. 1903-12-31.

Wentworth, E.N., Remick, B.L., 1916. Some breeding properties of the generalized Mendelian population. Genetics 1, 608–616.

Yates, F., 1952. George Udny Yule. 1871–1951. Obit. Not. Fellows R. Soc. 8, 308–323. https://doi.org/10.1098/rsbm.1952.0020.

Yule, G.U., 1902a. Mendel's laws and their probable relations to intra-racial heredity [part 1]. New Phytol. 1, 193–207. https://doi.org/10.1111/j.1469-8137.1902.tb06590.x.

Yule, G.U., 1902b. Mendel's laws and their probable relations to intra-racial heredity [part 2]. New Phytol. 1, 222–238. https://doi.org/10.1111/j.1469-8137.1902.tb07336.x.

Yule, G.U., 1906. On the theory of inheritance of quantitative compound characters on the basis of Mendel's laws—a preliminary note. In: Wilks, W. (Ed.), Report of the Third International Conference on Genetics. Spottiswoode & Co, London, pp. 140–142.

What natural selection must be doing: R. A. Fisher's early synthesis

> *Personally, I am* sure *that we ought to go on guessing, as intelligently as may be, and if it is an error it seems one on the generous side to do some of it in public.*
> **Fisher, letter to Leonard Darwin, October 11, 1930.**

As I noted in the introduction, one of the main arguments of this book is that the program pursued by the biometricians and the often quite subtle and sophisticated philosophy of science that grounded it are interesting not merely for their own sake, though I hope that by now the reader stands convinced of that claim. More than this, there is a genuine sense in which the kind of work that we saw in the last chapter forms a connecting link via which those ideas—the biometrical world-view as we have seen it elaborated above—were transmitted into the heart of early work on the Modern Synthesis. Far from being an irrelevant blind alley in the history of biology, then, biometry informed the approach to evolutionary theory that would go on to shape the majority of work on evolution throughout the 20th century.

There are a myriad of different connections to which one might point in order to make this claim. A handful of them are worth mentioning briefly as a glimpse of this diversity, drawing only from the collection of authors that I discussed in the last chapter. As A. W. F. Edwards has noted, Lock's textbook of evolutionary biology was used in lectures by E. B. Wilson at Columbia, and these lectures convinced H. J. Muller to continue in his study of genetics; Sewall Wright also read the work during graduate school (Edwards, 2013, p. 529). Thomas Hunt Morgan's research group, including Alfred Henry Sturtevant and Calvin Bridges, also read and discussed the book at length (Edwards, 2013, p. 533). E. S. Goodrich was a pivotal player in the Oxford biological community for decades, which connected him intimately with E. B. Poulton and the Huxleys, and he was funded for a number of years by James Mark Baldwin, which ties his work to the United States as well (Morrell, 1997, chap. 7). J. B. S. Haldane's only formal training in biology came from attending Goodrich's lectures at Oxford (Dronamraju, 2017, p. 13). Both Goodrich and Thomson's textbooks, as Peter Bowler has expertly contextualized them within the tradition of early-20th-century popular science, became known as models for pedagogical clarity and the popularization of biology (Bowler, 2009).

The Rise of Chance in Evolutionary Theory. https://doi.org/10.1016/B978-0-323-91291-4.00003-0

In short, then, the works that I discussed in the last chapter were widely read, and we thus have quite strong evidence that they would have been known by a variety of authors in the early days of the Modern Synthesis. How might the biometrical perspective that I have developed over the last few chapters actually look when expressed in the theoretical contributions of one of these biologists? To that end, I will turn to a detailed analysis of the work of the Synthesis "architect" most closely linked with questions of chance and statistics in evolution.

In 1909, a strange undergraduate named Ronald Aylmer Fisher arrived at Gonville and Caius College, Cambridge (Fig. 6.1). In a rather paradoxical manner, we simultaneously know quite a lot about Fisher's ideas as they developed even during his youngest years, and yet lack much of the information that could tell us what kinds of sources he was bringing together to form the new approach to natural selection, eugenics, and mathematical statistics that he would begin presenting to his fellow undergraduates just a few years later. Fisher's sole autobiographical reminiscence about this period, almost four decades later, is as short and unilluminating as it is untrustworthy:

FIG. 6.1

R. A. Fisher, 1913.

Credit: R. A. Fisher Archive, University of Adelaide, Rare Books and Special Collections
(http://hdl.handle.net/2440/81675).

I first came to Cambridge in 1909, the year in which the centenary of Darwin's birth and the jubilee of the publication of The Origin of Species *were being celebrated. The new school of geneticists using Mendel's laws of inheritance was full of activity and confidence, and the shops were full of books good and bad from which one could see how completely many writers of this movement believed that Darwin's position had been discredited.*

(Fisher, 1947)

Even the most cursory reading of the surviving text of two lectures Fisher delivered at undergraduate eugenics society meetings in 1911 and 1912 shows us that Fisher very quickly came to a wide-ranging and rather idiosyncratic understanding of natural selection, eugenics, philosophy, and Anglican faith, the interactions of which remain a puzzle for those interested in tracing the foundations of Fisher's later evolutionary theory. Our first task, then, is to investigate the various threads upon which Fisher was drawing in order to fashion his early vision of evolution by natural selection.

Fisher's sources

Let's begin with what we know about the young Fisher's grounding in evolution. In his last year in secondary school, and despite the fact that he had decided to opt for a university scholarship in mathematics rather than natural sciences, Fisher chose as a school prize the collected works of Charles Darwin, which accompanied him as he set off for Cambridge (Box, 1978, p. 17). For the rest of his life, he held Darwin as one of his scientific heroes, a sentiment which was only underlined and strengthened by his long-running friendship with Leonard Darwin, built in the context of the Eugenics Society. Biographers and historians have not hesitated to refer to him as the unofficial grandson of Darwin (via the childless Leonard), and this seems to be a justified ascription. During his first several years at Cambridge, then, he was already an ardent devotee of a gradualist, Darwinian understanding of natural selection—not an easy position to hold in a Cambridge that was dominated by the equally ardent Mendelism of Bateson and Punnett.

We can be absolutely certain that Bateson's 1909 textbook (discussed in the last chapter) served as his introduction to Mendelism—while Bateson and Fisher only overlapped at Cambridge for one year, before Bateson left his newly created Professorship of Genetics to take over the direction of the John Innes Horticultural Institute in 1910 (Box, 1978, p. 22). The two men probably never met, but Punnett filled the professorship after Bateson departed, and as we have already seen, there was little daylight between Bateson and Punnett. (Fisher would become the third person to occupy the chair, in 1943.) It is equally reasonable to assume that Fisher probably owned a copy of Punnett's own textbook, *Mendelism*, the second edition of which had just appeared (Punnett, 1907).

With respect to his exposure to the biometrical school, we know that by his earliest discussions of evolutionary theory in 1911, he was already intimately acquainted with the debate between Pearson and Yule over natural selection, correlation, and dominance.

He had therefore quite likely read much of Pearson's "Mathematical Contributions" series of journal articles (and he exhibited familiarity and contempt in equal measure for Pearson's method of curve-fitting by moments, developed in those articles, for the rest of his career; see Fisher, 1937). He even followed the discussion of selection and correlation into a number of its later, more obscure tributaries (Brownlee, 1910; Snow, 1910), as we will discuss shortly in connection with his first publication on biometry and Mendelism.

We also have strong documentary evidence, thanks to Edwards, that Fisher read the copy of the second edition of Lock's *Recent Progress in the Study of Variation, Heredity, and Evolution* shelved in the Gonville and Caius College library—a graph found within it bears an axis label in Fisher's unique handwriting (Edwards, 2013). This would have put Fisher directly in touch with the kind of integrative, cumulative approach to evolution upon which I have placed such importance. As he wrote to Leonard Darwin during the preparation of the *Genetical Theory* in 1929, reading these works forced one to question the dominance of the Mendelian school; he mused, rhetorically, "how far did [Bateson] alienate the better biologists, e.g. Poulton, Goodrich from Genetics, and how much did it matter?" (R.A.F. to L.D., January 21, 1929; Bennett, 1983, p. 96). It therefore seems to involve no great inferential leap to place Fisher squarely within the biological current that I have illustrated here.

This constitutes all the available evidence concerning Fisher's reading in evolution. We should now turn to a few other important elements of his thought that are essential for forming a complete picture of his work. First and foremost among them is eugenics. As I noted in the introduction, to genuinely focus on the eugenic beliefs of the various authors that I have treated throughout this book would be the subject of an entirely different (and much longer) work, and indeed one that has largely already been written several times over. But as authors like Pauline Mazumdar and James Moore have noted, Fisher's commitment to eugenics forms a widely discussed and inarguably central element of his work as a biologist. Perhaps most provocatively, Donald MacKenzie (1981) has insisted that it is eugenics that is entirely in the driver's seat for Fisher's evolutionary and statistical work—his work on natural selection, mathematical statistics, experimental design, and more can and must all be understood through the lens of this pursuit. Such a radical conclusion is, it now seems, too hasty. Fisher's analysis of variance approach (ANOVA), for instance—one of his most lasting contributions to mathematical statistics and experimental design—can be conclusively traced to his work at Rothamsted Experimental Station on agricultural experiments. As Giuditta Parolini writes, "primary and secondary sources offer more convincing evidence that the development of the analysis of variance took place during Fisher's work at Rothamsted," as a "response to the experimental problems posed by the research done" there (Parolini, 2015a, pp. 317–318, 303).

In short, it seems that Fisher's eugenic beliefs are as problematic for us as interpreters as the other ingredients in his intellectual biography. While he was a cofounder of the undergraduate Cambridge Eugenics Society, a member of a group of friends with a penchant for quoting Nietzsche at one another and giving each other names from Norse mythology, and would spend several decades as a professional acting in various capacities for the Eugenics Society, until its lack of scientific rigor finally drove him away, he was also undeniably a bizarre sort of eugenicist, in whose works, as Mazumdar puts it, "the tone of frightened loathing of the lower class which is so plain in the writing of most of the eugenists is not so blatant" (Mazumdar, 1992, p. 102). We thus have no choice but to treat this as yet another element to be disentangled in Fisher's writings; no easy hypothesis of all-sufficiency is here to be found.

A second, long-standing influence for Fisher is his Christianity, as explicated masterfully by James Moore (2007) and John Turner (1985). Fisher's faith was acquired honestly, via his devout mother, and cemented by shock, after she died at the age of 49, when Fisher was only 14 (Box, 1978, p. 14). His wife's family, a famous clan of low-church evangelist missionaries, practiced their faith in a quite different way than the Anglican Fishers. In turn, eugenics was to take on a near-religious aspect for Ron (as he was known to his family) and his wife, as they took up subsistence farming (of fruits, vegetables, pigs, chickens, and children) in the spirit of genuine eugenic practice, after Fisher was rejected for military service in World War I due to his poor eyesight (another inheritance from his mother).

How did Fisher's Christianity impact his biology? I will return to the subject in the last section of this chapter, but for now, we may underline a focus on activity, creation, and free will. Faith in a deterministic universe (and, by extension, a harsh, Calvinist predestination) was no faith at all for Fisher. His evolutionary theory would, therefore, focus on the creative, open, free interaction between organisms and the world. "In this interaction," as Turner aptly puts it (quoting Fisher from 1950), "'living things themselves are the chief instruments of the Creative activity,' for their decisions, freely made, affect their probabilities of survival, death, and reproduction, which are themselves the agencies of natural selection" (Turner, 1985, pp. 161–162).

To close this discussion of Fisher's influences, two properly philosophical commitments should be briefly mentioned—explicating them in full is a project to which I will return below. First is a deep commitment to stochastic scientific theorizing. Jon Hodge has especially emphasized that, along with Darwin, Fisher's other scientific hero was Boltzmann, whose thermodynamics was introduced to him in a post-graduation year spent working with the physicist James Jeans in Cambridge (Hodge, 1992, p. 241).[a] Fisher's appreciation of statistical physical theory dovetailed perfectly with the second philosophical commitment that I want to quickly highlight here—to indeterministic causation. With statistical mechanics having, he thinks, conclusively demonstrated the impossibility of a thoroughgoing Newtonian, deterministic picture of the world, Fisher will argue that a generalization of causation to encompass probabilistic influence is absolutely essential to capture the nature and structure of modern science, including the very evolutionary theory that he would craft in the coming decades.

There we have it, then: as Fisher takes up dedicated work on evolutionary biology near the end of the 1910s, he is approaching it from a perspective that included

---

[a] Unfortunately, there is very little documentary evidence concerning what Fisher studied or took away from this time working in physics.

biometry, Mendelism, and the attempts throughout the previous decade to reconcile the two, as well as eugenics, Anglicanism, an internship in statistical mechanics, and a philosophical toolbox containing no less than a radical revision of the Newtonian approach to scientific theory. How are we supposed to see these pieces as coming together? Of course, we might not: Depew and Weber have described this mixture (adding to my list Nietzsche and British imperialism) as "an extraordinary, and perilously incoherent, compound" (Depew and Weber, 1995, p. 249). My sympathies lie rather with Moore, who, drawing on Hodge's reconstruction, hopes to summon the "historical imagination" required for us to integrate these seemingly disparate threads (2007, p. 115). To do so, I want to look at three milestones in Fisher's corpus: first, his earliest work, comprised of his 1918 and 1922 papers, then a brief interlude on his innovation in statistics and experimental design, and finally the *Genetical Theory of Natural Selection* (hereafter *GTNS* for short), one of the landmark works of the nascent evolutionary synthesis, published in 1930. I will then turn to the thornier task of making sense of Fisher's handling of indeterminism, statistical physics, free will, scientific theories, and "creativity"—the real place, I will argue, where Fisher makes space for the biological theorizing developed by the biometricians to become part of the permanent furniture of evolutionary biology.

I should also note here that my story will, somewhat abruptly, stop in 1930. I will say more about this choice in the next and final chapter, but, in short: if my aim here has been to detail the intellectual development and transmission of a certain approach to the life sciences, and we can demonstrate that such an approach is to be found in *GTNS*, then there can be no doubt that such ideas were there for the taking in the creation of the Modern Synthesis. We can turn to other, more capable historians of the Synthesis period and the development of molecular biology for the tale from 1930 to today.

The early years

One remarkable feature of Fisher's work upon first contact with it is its consistency. Beginning even with lectures that he delivered as an undergraduate (reprinted in Bennett, 1983), through his groundbreaking early work, and into the books of his mature career, his approach to and understanding of evolutionary theory is impressively stable—all the more so when we consider that this spanned the massive scientific changes that were wrought in the life sciences from 1911 to the 1950s. That said, I want to be cautious here. As Hodge has rightly warned us, any effort to understand the broad scope of Fisher's work will have to make sure that it accounts for "how his Mendelism expanded from being merely a theory of factors subject to linkage and segregation and so on, to being a theory of genes" in the modern sense of the term, as well as "how his selectionism was developed in contrasting gene frequency changes…due to selection and those due to such chance accidents…as would eventually, in later years, be dubbed random genetic drift" (Hodge, 1992, p. 239). In short, two major external events would indeed have a profound impact upon Fisher's early evolutionary theorizing: his increasing engagement with Sewall Wright (and, a fortiori, with genetic drift via Wright's shifting balance theory), and the progress in molecular genetics produced most notably by T. H. Morgan's group in the United States. It is, therefore, profitable for my purposes to divide Fisher's work into two periods: the articles of 1918 and 1922 which, by and large, did not engage with such questions, and the *Genetical Theory of Natural Selection* itself, which did.

It is something of a tradition in the history of biology to mark Fisher's 1918 paper, "The correlation between relatives on the supposition of Mendelian inheritance," as the first genuine work of the Modern Synthesis. The discussion in the last chapter, I hope, has offered ample evidence as to why I believe such an accolade isn't worth very much—for there is more continuity here than is commonly suspected. Granted, Fisher is clearly a brilliant scholar, and his work combines a level of refined mathematical acumen with a familiarity with biological context theretofore unmatched. Unfortunately, Fisher is infamously difficult to read. Nearly every review of his work mentions this fact at least a few times in passing, even the positive ones, and it took a 64-page annotated reprint, published in 1966, which corrected dozens of typos and mistakes throughout many parts of the 1918 article, to present its ideas to a contemporary professional audience (Moran and Smith, 1966).

We should start by backing up a bit, and elaborating somewhat on the argument between Pearson and Yule over inheritance and dominance. In what he took (for the rest of his life!) to be a fatal criticism of Mendelism, Pearson in 1904 had derived that, on the Mendelian assumption of complete dominance, the correlation between descendants and ancestors should be precisely one third—that is, one third of the variability of offspring from the average value in the population should be explained by the variability of their parents from that same average (Pearson, 1904). Unfortunately for Mendelism, Pearson noted, a number of large-scale empirical studies undertaken at the Galton Laboratory had conclusively demonstrated correlation coefficients in the real world much closer to one half. So much for the ability of Mendelism to explain the array of characters that had thus been analyzed, which encompassed everything from basset-hound coat color to human eye color and height. Yule responded several years later by questioning Pearson's assumption of complete dominance—that is, that the recessive trait would be completely absent from the phenotype in hybrid organisms (Yule, 1906). If we make room for the possibility of partial dominance, it is easy to derive correlation coefficients closer to one half—indeed, one can interpret the correlation coefficient as an estimation of the extent of dominance, instead of the other way around.

In the interim since 1906, several biologists had taken note of this debate and had realized that it could profitably be digested into three important questions, each of which bears on the way in which we should take advantage of the biometricians' wealth of accumulated data about variation in natural populations. First, what exactly *is* a correlation coefficient, and how should it be measured? A number of seemingly technical choices in calculation, it was discovered, could, in fact, have a material impact on the values obtained, as well as on the relationship that those values held to the objects they were taken to be measuring. Second, what kind of correlation coefficients should we actually expect to observe between various kinds of relatives, in various kinds of Mendelian populations? This question forms an extension of Pearson and Yule's debate, in the same direction as much of the work discussed in the last chapter—deriving the theoretically predicted correlations for different relatives (parents, siblings, cousins, and so forth) given different assumptions about underlying character structure and transmission (degrees of dominance, mating regimes, etc.). And lastly, where does the influence of the environment go? If the correlation coefficient tells us that a given fraction of variability is explained by inheritance from ancestry, where exactly did the rest come from? Some of it is certainly due to random variability between parents and offspring, the sort of thing that Darwin took to be the grist for natural selection's mill. Some of it is probably the result of environmental variability as well, and some of it might be wrapped up in correlations between parental environment and offspring environment. Could these factors be disentangled, and if so, how?

In introducing these problems in his 1918 paper, Fisher of course mentions Pearson and Yule's debate directly—in fact, Leonard Darwin had written to him while the paper was in preparation to ask "whether it would not be best to write out your remarks in their final form, and submit them to Yule, or get us [i.e., the Royal Society] to do so" (L.D. to R.A.F., September 3, 1915; Bennett, 1983, p. 65). Fisher also invokes two other statisticians who had been working on solutions to these questions: the epidemiologist John Brownlee, who had published an analysis of correlation coefficients in the *Proceedings of the Royal Society of Edinburgh* (Brownlee, 1910), and a theoretical elaboration of Mendelian inheritance prepared by E. C. Snow, who had been trained directly by Pearson in the Biometric Laboratory (Pearson, 1938, p. 182), and whose paper was communicated to the Royal Society by Pearson himself (Snow, 1910). One would be correct in assuming that this meant that Snow's paper largely agreed with Pearson's calculation of the Mendelian correlation coefficients.

It is important to stress the extent to which this was already a highly sophisticated debate, even as early as 1910, despite its rare appearance in the secondary literature. Brownlee, after introducing the difference between the values of parent-offspring correlation derived by Pearson and Yule in much the way I have above, notes that this apparent contradiction entails that "some means of reconciling theory and observation must be found" (Brownlee, 1910, p. 476), and goes on to explore a number of explanations that might do the job, including not just the technical means of calculating the correlation coefficient, but also assortative mating, parental selection, the presence of more than two alleles per locus, and a discussion of the values of fraternal correlations, nearly all of which are also present in Fisher's paper.

Into this somewhat crowded field enters Fisher, with the stated aim of generalizing upon Yule's results in the direction already taken by Brownlee and Snow. As he introduces the problem,

> *Several attempts have already been made to interpret the well-established results of biometry in accordance with the Mendelian scheme of inheritance. It is here attempted to ascertain the biometrical properties of a population of a more general type than has hitherto been examined, inheritance in which follows this [i.e., the Mendelian] scheme.*

(Fisher, 1918, p. 399)

In this emphasis on generalization, we see yet another feature which manifested itself as something like a self-acknowledged tension throughout Fisher's work. Not for the last time will we see Fisher attempting to strike a golden mean between what to him was the obvious desirability of highly general, mathematical theory, and the attention to detail and contingency that was necessary for understanding the biological world.

He then turns to two theoretical tools in order to demonstrate the compatibility of biometry and Mendelism. The first is already familiar. If we wish to analyze continuous variation, Fisher writes, "the simplest hypothesis" concerning the underlying system of inheritance, given our current state of knowledge, "is that such features as stature are determined by a large number of Mendelian factors, and that the large variance among children of the same parents is due to…segregation of those factors in respect to which the parents are heterozygous" (Fisher, 1918, p. 400). Fisher thus takes as a starting point the results of work like East's, which had already shown that the appearance of continuous variation could be produced by a large number of simple Mendelian characters.

What had not been done before, however, was Fisher's next step, which was to analyze the variance of the population of offspring in an effort to determine its sources. A bit of caution is required here: Fisher would later use the phrase "analysis of variance" to refer to a very particular method (now abbreviated as ANOVA) of processing the results of statistical experiments and of visualizing that result in a tabular form. This procedure was developed in detail only in the next decade as Fisher worked on the results of agricultural experiments at the Rothamsted Experimental Station, so by "analysis of variance" I do not (and could not) mean this formalized, well known analytic tool. What Fisher is doing, nonetheless, is analyzing the variance of the offspring population—there are no other words to describe it, Fisher himself having had to coin the word "variance" in this very paper as a way to talk about a quantity that had not previously been important to statistical analysis. He uses this tool to take up one of the challenges we discussed above. Having started with the population as built from a large number of Mendelian factors, he writes, "we will attempt to determine how much of the variance, in different measurable features, beyond that which is indicated by the fraternal correlation, is due to innate and heritable factors" (Fisher, 1918, p. 400). Put more clearly, if we measure the correlation between siblings in a number of different characteristics, we find that its value is about 0.54, or that 54% of the variance in siblings is accounted for by direct inheritance, that is, by ancestry. Setting that fraction aside, we need to figure out how much of the remaining 46% is due to processes like random Mendelian segregation (also a feature of "innate and heritable factors," in Fisher's terms), and how much is due to the environment. Following on the prior work of Galton and Pearson (and here, one might well argue, enters support from his eugenic emphasis on the potency of the innate over the acquired), Fisher is a priori rather convinced that the environment will not have a substantial impact; we thus have quite a theoretical task in front of us in order to explain a full 46% of variance in some other way. The most likely candidate is the effect of Mendelian dominance.

Putting aside the details of an argument that becomes very technical very quickly, Fisher's major insight here is to realize that dominance would have a peculiar effect on the correlation coefficients between siblings, because they will only vary from one another in characters for which their parents were hybrids. This gives us just the mathematical leverage that we need "to distinguish, as far as the accuracy of the existing figures [i.e., Pearson's data on correlations in humans] allows, between the random external effects of environment and those of dominance" (Fisher, 1918, p. 406). He goes on to do this, taking into account a number of different challenging population structures, including cases of assortative mating, multiple allelomorphism, and linkage—noting in passing that the Law of Ancestral Heredity falls out as a special case when a correlation between parents is present due to correlation between their observed, "biometrical" trait values (Fisher, 1918, pp. 420–421).

In sum, he concludes the paper by noting that he has successfully "distinguish[ed] dominance from all non-genetic causes, such as environment," and explained a puzzling, persistently high value for sibling correlations, which "is itself evidence in favor of the hypothesis of cumulative [Mendelian] factors" (Fisher, 1918, p. 433). All our best data (excepting a single strange value from Pearson's group concerning first cousins) fits such a view quite well, and further "shows that there is little or no indication of non-genetic [i.e., environmental] causes" (Fisher, 1918, p. 433).

Here, then, is the foundation Fisher needs to build his theoretical edifice. There is no incompatibility between the sheaves of population data obtained by the biometricians (and, one hastens again to add, useful for Fisher's eugenic pursuits) and the Mendelian theory of inheritance. We have saved the phenomena. Now it is time to turn to evolution.

Before we follow Fisher in doing so, however, it is worth pausing to say something about the circumstances surrounding the appearance of this 1918 paper. At the encouragement of Leonard Darwin (and, as noted above, possibly with the approval of Yule), Fisher had it communicated in late 1916 to the *Proceedings of the Royal Society*. In a stroke of amazingly bad luck, it was assigned Punnett and Pearson as its two peer reviewers (Box, 1978, p. 59)—a pairing that could only have been made more polemical if Bateson had replaced Punnett. Rather unsurprisingly, neither liked it: it was too mathematical and non-biological for Punnett (who would for decades do his best to describe Fisher as an unwelcome statistician moonlighting in biology; see Punnett, 1930), who thought that Pearson might like it, and it made far too many concessions to Mendelism for Pearson, who thought that Punnett might like it. With the support of J. Arthur Thomson, however, it was moved to the *Proceedings of the Royal Society of Edinburgh* and published there. Frankly, and despite its current fame, it is unclear to what extent it was actually read by fellow biologists at the time, and to the extent it was read, it is unclear whether it was actually understood. This is a persistent worry with all of Fisher's early work, as his methodology is so unique and complex that prior to his 1928 article, which generated quite a bit of discussion in *American Naturalist*, it is difficult to trace Fisher's immediate impact.

Returning to his evolutionary theorizing, it was not only the correlations between relatives that would be found to yield to these new analytic tools. A few years later, in 1922, Fisher began to use these methods to look into population change. What he proceeded to lay out was a set of claims, now relatively familiar to students of evolutionary biology, about the dynamics of the introduction of novel genes into populations. When a gene is first introduced by mutation, "its survival depends on that of the individual in which it occurs, and this chance is variable from species to species"—or put more provocatively, "while it is rare, its survival will be at the mercy of chance, even if it is well fitted to survive" (Fisher, 1922a, p. 326). There is no need for despair, however, as while, to take an example of Fisher's own, only around 2% of new mutations in one of his models would survive for 100 generations, those that do will on average be represented in half the members of the population. Fisher next turns his fire to genetic drift, arguing that its influence must be essentially non-existent, as the amount of time that would be required for it to act would be much longer than the average time between environmental changes, and hence the shuffling of (more powerful) selection pressures. Looking at the impact of selection, Fisher derives that "even the slightest selection is in large populations of more influence in keeping variability in check than random survival" (Fisher, 1922a, p. 334).

Consider finally the dynamics of selection itself. Because the process of increasing fitness via random mutation penalizes large jumps—climbing a hillside by movement in random directions is best done in small steps, as the hill curves away from you horizontally around as well as behind, and large jumps are thus likely to end up lower than they started—we should expect "that the large and easily recognized factors in natural organisms will be of little adaptive importance, and that the factors affecting important adaptations will be individually of very slight effect" (Fisher, 1922a, p. 334). Natural selection works best when it can sift through a very large number of mutations, each of which individually might have little consequence, but which will accumulate over time into real, directional population change.

By 1922, then, Fisher has already constructed a picture of natural selection with the following features, all of which would remain as elements of his later work on evolution. First, he has made room for the saving of the biometrical phenomena—he has both demonstrated that the vast extant trove of biometrical data is consistent with our growing understanding of the mechanism of inheritance, and that the impact of dominance is relatively easy to account for in modeling and yields empirically plausible results. Evolution by natural selection, then, will be supplied with the raw material of mutations, again at a reasonable rate of mutation. These mutations' initial fates will be governed by chance, when their survival is intimately causally tied to the survival of the organisms which carry them, until (as will eventually occur with at least some mutations) they are carried by enough organisms for their fate to be determined by natural selection. Indeed, this is the most efficient way for selection to work, with a constantly provided flux of mutations of individually small contribution. Such a process provides—as a matter of mathematical consequence, for Fisher—little room for drift or neutral evolution to operate, as most of the relevant parameters for drift's effect scale inversely with population size, which will, Fisher thinks, be very large in most natural cases.

At the risk of repeating the refrain of this and the prior chapter too many times, the continuity with the prior goals of the biometricians and their descendants in the interregnum period seems clear enough. Fisher's theory is a *statistical approach* to the *action of natural selection* in *generational time*, at the *population level*, harmonized with *Mendelian transmission*. Fisher has already found the Holy Grail for which Weldon was searching in 1905.

Interlude: Fisher at Rothamsted

Before we move on to Fisher's development of this perspective in his late-1920s works and *GTNS*, however, we should stop to reflect at least briefly on the relevance for the argument here of the major advances he was simultaneously making in statistical methodology and experimental design. Having failed to secure academic work after his time at Cambridge, Fisher spent the war years as a secondary mathematics instructor—a job which he absolutely detested, and from which he resigned the moment the war ended (Box, 1978, p. 60). Having declined Pearson's offer of a research post at the Galton Laboratory (fearing, almost certainly with good reason, that he would only be allowed to produce work of which Pearson approved; see L.D. to R.A.F., August 7, 1919; Bennett, 1983, p. 90), he finally found his way to a position as statistician at the Rothamsted Experimental Station, an agricultural research facility which had made a name for itself as a home for research into fertilizers. Fisher arrived in 1919, having been summoned as someone who could, in the words of the station's director, "examine our data and elicit further information that we had missed" (Box, 1978, p. 96), particularly with respect to their long-term experiments with wheat (Fig. 6.2)—data dating back to the middle of the 19th century that had in many cases yet to be systematically analyzed (Parolini, 2015a, pp. 304–305). Fisher would remain there until he succeeded Karl Pearson as the Galton Professor of Eugenics at University College in 1933.

To briefly describe Fisher's pursuits during these years is not an easy task. They included, of course, the drafting and publication of the 1922 paper discussed in the last section, as well as *GTNS*, although Fisher's evolutionary work (including even some Mendelian breeding experiments he carried out at home) was described by his biographer and daughter Joan Fisher Box as forming part of "his extraprofessional activities" (Box, 1978, p. 167). His focus during this period was rather on the statistics of small samples and the design of experiments, both prepared in close contact with the field work and analysis tasks that were constantly taking place at Rothamsted. Reasons of space prevent me from doing justice to either of these parts of Fisher's corpus, but a short summary will at least help paint a picture of the breadth and importance of his efforts during this period.

As far as formal tools are concerned, the main question with which Fisher concerned himself was our ability to make statistical inferences from the results of small experiments. Much of the work of previous investigators like Pearson had been targeted at large-scale, population-level inferences, where the properties of a

FIG. 6.2

The Broadbalk long-term wheat experiment at Rothamsted Experimental Station, one of the agricultural experiments which Fisher was hired to analyze.

Credit: Macdonald, A.J. (Ed.), 2018. Guide to the Classical and Other Long-Term Experiments, Datasets, and Sample Archive. Rothamsted Research, Harpenden, UK, cover. https://doi.org/10.23637/ ROTHAMSTED-LONG-TERM-EXPERIMENTS-GUIDE-2018.

distribution can be estimated from the curve derived directly from empirical observation. But what if, as was often the case at Rothamsted, we don't have this much data? How ought we to estimate the effect of, say, a manure treatment, if it has only been replicated a few times in one of the experimental plots whose data Fisher was now expected to analyze? Fisher was the first to build many of the tools that are needed to tackle such questions. Over the course of this quarter-century, he originated the maximum likelihood approach to parameter estimation, popularized and rigorized the use of William Gosset's (or "Student's") *t*-test of statistical significance, clarified some persistent mistakes in the application of Pearson's chi-squared test, and developed his own significance test, now known as the *F*-test (named for Fisher). He demonstrated not only the applicability and usefulness of these various methods, but was unusually adept at showing the ways in which they related to one another. (Box ascribes this talent to Fisher's equally unusual ability to visualize high-dimensional geometry, perhaps, she argues, a skill he was forced to learn as a result of his poor eyesight.) His 1925 textbook *Statistical Methods for Research Workers* was reprinted for decades, and included thorough explanations of all of these significance tests, as well as correlation coefficients, the development of the now ubiquitous *p*-value, and a presentation of the method of the analysis of variance (ANOVA) which he had finally clarified and formalized in his studies of agricultural data.

As regards experimental technique (which would eventually appear in the synthetic 1935 book *The Design of Experiments*), Fisher worked on the generalization of the methods of patch-based experimental design, which were currently in common use at Rothamsted (Parolini, 2015a, b). Via advanced work on Latin square experimental designs, in which a field is subdivided into blocks that receive differing treatments in an effort to average over the heterogeneity in the underlying conditions (here, the soil and micro-climate), he began to advocate the use of experimental randomization, and eventually developed the now common factorial approach to experiment design, which allowed for a single trial to discriminate between a vast number of different influences on its results. Field workers could choose the particular influences in which they were interested, neglecting others and allowing them to be mixed in with the error that resulted from natural variability (confounded), all with much more compact plots than would have been possible using prior methods.

In short, Fisher's perspective on evolution and selection during this period was richly informed not only by academic engagement, but by the kinds of everyday analytic and experimental problems that confronted him in his work at Rothamsted. It will be instructive, then, to be on the lookout for ways in which signs of this work might be found within Fisher's evolutionary theorizing.[b]

The Genetical Theory of Natural Selection

Practically since they began corresponding in the mid-1910s, Leonard Darwin had been urging Fisher to publish. Fisher had started preparing a manuscript for a first book focusing mainly on eugenics and human inheritance, but Darwin was unconvinced. "It is worth taking great pains with your first book," he wrote Fisher, "even though a book is an awful grind" (L.D. to R.A.F., August 22, 1919; Bennett, 1983, p. 70). As Fisher's work on a statistical approach to natural selection had begun taking shape over the 1920s, however, Darwin's tone changed. "I look on my letters to you," he wrote in late 1927, "in the light of pins, the pin pricks to urge you on with your great work on the mathematical theory of inheritance!" (L.D. to R.A.F., November 1, 1927; Bennett, 1983, p. 82).

By this point, Fisher had already begun to tie up a number of loose ends in his presentations of 1918 and 1922. In 1927, he took on one of the Mendelians' favorite

---

[b] Provocatively, for instance, Mazumdar locates in precisely this connection a significant point of friction between Fisher and other members of the Eugenics Society over their approach to human inheritance (Mazumdar, 1992, pp. 128–130).

examples—the evolution of mimicry in butterflies. As we saw in Chapter 5, mimicry was the context in which Punnett had reproduced, to no great utility, Norton's mathematical modeling of the rate of natural selection (Punnett, 1915). Now, however, Fisher could confidently write (with Punnett and Bateson the intended targets) that "the bearing of genetical discoveries, and in particular of the Mendelian scheme of inheritance, upon evolutionary theory is quite other than that which the pioneers of Mendelism originally took it to be" (Fisher, 1927, p. 269). He saw two open questions as pertinent, tackling them both in an article the following year (Fisher, 1928a). His new approach developed in the context of mimicry—that some genes served as switch-like modifiers that could control the expression of other genes, which would then be free to mutate without a loss of fitness if they were "switched off"—was particularly applicable to the solution of these problems. Fisher argues that because the vast majority of mutations will be disadvantageous compared to the wild type, a new mutation will be expressed in a population almost exclusively as a heterozygote. A quick derivation confirms, then, that we really need consider the effectiveness of selection only on those heterozygotes, for this type "will be so enormously more frequent than the homozygote that any selection of modifiers which is in progress will be determined by the reaction of the heterozygote" (Fisher, 1928a, p. 120).

We then have two moves available to us. First, Fisher notes that this process could offer an explanation for the evolution of dominance itself as a property of some alleles—and particularly, as he had noted elsewhere, the somewhat strange fact that dominance is a common property of wild-type alleles but a rare property of many other commonly observed mutations. As he puts it,

Whatever were the relative frequencies of dominants, recessive[s] and intermediates of various grades among mutations at their first occurrence, we should expect, if the above selective process has had time to produce great effects, to find that the greater number of recurrent mutations had become completely recessive.

(Fisher, 1928a, p. 122)

Why is this the case? The argument runs roughly as follows. He has already claimed that when natural selection acts upon most mutations, since the majority are disadvantageous, it will effectively be acting only upon the hybrids, and almost negligibly upon organisms that are homozygous for the new mutation. The optimal state for the mutation, then, and the one that we would expect if it is relatively common (and thus selection has had many chances to act upon the modifiers which control it), would be for the heterozygote to do *nothing at all*—to simply express the wild type. In short: we would expect that common mutations will evolve into being completely recessive, just as they in fact have done for the majority of the mutations Morgan and his team had detected in fruit flies. We thus have an explanation for the evolution of dominance and recessiveness.

Finally, such a structure still leaves room for us to understand the long-term effects of the kind of slow, mass natural selection that he had considered in 1922. One might have been worried, Fisher writes, that the dominance which appears to be present in the vast majority of wild-type genes would indicate that they are "clearly of a different nature from the mutant genes which arise from them." If this were so, and there were really two dissimilar categories of genes, we would no longer have a clearly defined, unitary process which produced mutations, filtered them through natural selection, and potentially replaced the former wild type. But "this difficulty will lose its force," he argues, "if it appears that there is a tendency always at work in nature which modifies the response of the organism to each mutant gene in such a way that the wild type tends to *become* dominant" (Fisher, 1928a, p. 116). It is precisely this process that Fisher takes himself to have described above. If dominance and recessiveness are simply about the reaction of the organism's overall collection of modifier genes to new mutations, then there is nothing standing in the way of the long-term effectiveness of the selection of small mutations to produce progressive, permanent change in the wild type, as these modifiers are also under the control of long-term selection. In short, if it is evolutionarily advantageous for a mutation to become dominant, natural selection is empowered to make it so.

As a brief aside, and in marked contrast to the 1918 and 1922 papers, it is worthy of note that a flurry of discussion followed the publication of the 1928 paper. Several responses immediately appeared in *American Naturalist*, and this paper was the catalyst for Fisher's (first amicable, eventually much less so) dialog with the American geneticist Sewall Wright (Fisher, 1928b, 1929; Wright, 1929). Several of the exchanges that Fisher was drawn into, particularly one on the proper way to interpret mutations in the coloration of poultry (which had been another of Bateson's favorite cases), would eventually make their way into *GTNS*.

He then set his sights on the *Genetical Theory of Natural Selection* manuscript. Fisher wrote the volume quickly, between October 1928 and June 1929, dictating the non-mathematical chapters to his wife (Box, 1978, p. 186). It is, in the end, a peculiar book, difficult to read and difficult for us to interpret historically. It consists of twelve chapters—the first a historical treatment of Darwin's use of blending inheritance, then four chapters describing the general outlines of Fisher's theory of natural selection, dominance, and variation, two chapters treating sexual selection and mimicry as special cases worthy of more detailed analysis, and five chapters detailing human heredity, social class, fertility, the decline and fall of human civilizations, and our ability to intervene in these processes eugenically. He writes that "the deductions respecting Man are strictly inseparable from the more general chapters, but have been placed together in a group," as part of his effort to make the book more readable (Fisher, 1930, p. x). Many of the connections between the eugenics and the evolutionary theorizing are, however, not drawn explicitly, and hence it is left to us to "recognize the consonance of the latter, eugenical chapters with Fisher's neo-Darwinism in the rest of the book and elsewhere" (Moore, 2007, p. 114).

He consciously presents the work as rather preliminary or provisional, by emphasizing two points in the preface. First, the book's opening words: "Natural Selection is not Evolution" (Fisher, 1930, p. vii). Fisher does not intend to be putting forward a comprehensive theory of evolution here—a position that might be unexpected given that this work would later regularly be lumped in with the other systematic treatises of evolutionary theory produced by the Modern Synthesis. As he elaborated the point later in a letter to Julian Huxley:

> *The importance which you and Haldane attach to [GTNS]…gives me much plea-sure, but not a little embarrassment, for if I had had so large an aim as to write an important book on Evolution, I should have had to attempt an account of very much work about which I am not really qualified to give a useful opinion. As it is there is surprisingly little in the whole book that would not stand if the world had been created in 4004 BCE, and my primary job is to try to give an account of what Natural Selection must be doing, even if it had never done anything of much account until now.*

(R.A.F. to J.S.H., May 6, 1930; Bennett, 1983, p. 222)

Fisher's aim, then, is to offer a generalized theory of the action of natural selec-tion, and nothing more.[c] Of course, this itself is no small feat—keeping in mind that Galton, Pearson, Weldon, and others had all failed to do so in the preceding decades.

The second way in which Fisher marks out *GTNS* as provisional is intimately re-lated to the first: insofar as this will be a difficult piece of theoretical work, Fisher is aware that he doesn't have much of an extant body of theory to draw upon. "It seems impossible," as he puts it, "that full justice should be done to the subject in this way, until there is built up a tradition of mathematical work devoted to biological prob-lems, comparable to the researches upon which a mathematical physicist can draw in the resolution of special difficulties" (Fisher, 1930, p. x).

Having already introduced most of the features of Fisher's understanding of evolu-tionary theory, I can dwell here mainly on those pieces of the argument which are novel in *GTNS*. He begins by offering a treatment of Darwin's works, drawing particularly upon the then-recently published *Sketch* and *Essay* (see Chapter 1) to present an ac-count of Darwin's view of blending inheritance. He underlines the potential swamping of beneficial variations, casting it in his statistical framework as the challenge that, if inheritance is truly blending and mating random, the variance available for natural se-lection to act upon will be halved in every generation, rapidly becoming insufficient to drive future selective change. Bizarrely, given the actual history of particulate theories of inheritance, he argues that "all the main characteristics of the Mendelian system flow from assumptions of particulate inheritance of the simplest character, and could have been deduced *a priori* had anyone conceived it possible that the laws of inheritance could really be simple and definite" (Fisher, 1930, p. 8). This is an especially odd as-sertion given that, at the very least, Galton and Weldon had tried to do exactly this and nonetheless produced non-Mendelian distributions. He takes the results of Johannsen to have conclusively demonstrated the absence of blending inheritance in nature (Fisher, 1930, p. 18), which solves the worry of its progressive loss of variance, and when this re-sult is combined with the extant work on the production of continuous variation by large numbers of Mendelian characters, we are left with no further problems for the potential

---

[c] My thanks to Alex Aylward for emphasizing this point and the corresponding letter.

action of natural selection. The objective here is thus largely the same as that of the 1918 paper, but presented in a new way, by drawing a historical contrast with blending inheritance. Once again, as it had in 1918 and 1922, the establishment of the structure of continuous variation as arising from a large number of Mendelian factors serves as the basis for his results about the process of natural selection.

In the second chapter, Fisher turns to the derivation of what he will call the Fundamental Theorem of Natural Selection, which constitutes the book's main theoretical advance. Before discussing just what the theorem claims and how it should be understood, it is worth stopping to ask what he hoped to use it to accomplish. While the details are somewhat challenging to interpret, and I will return to the relationship between Fisher's work and statistical physics in the next section, the comparison that he makes between the Fundamental Theorem and the second law of thermodynamics is instructive. He writes that they are alike in being "properties of populations, or aggregates, true irrespective of the units which compose them" (Fisher, 1930, p. 36), in being statistical, and in defining change in terms of the global increase of a particular quantity (whether fitness or entropy). All of this seems to be consistent with another of Fisher's broad goals in *GTNS*—the pursuit of generalized theory. The Fundamental Theorem gives us a way to think about natural selection independent of particular characters and particular populations—a way for Fisher to demonstrate, as he wrote to Huxley, what natural selection *must* be doing as a matter of mathematical structure and logical consequence. It also clarifies the sense in which natural selection can be seen as "chancy," which Fisher considers here via "the objection which has been made, that the principle of Natural Selection depends on a succession of favorable chances" (Fisher, 1930, p. 37). This complaint is unfortunately still familiar today in the writings of creationists, who decry selection as requiring the conjunction of numerous low-probability events, the conjunction becoming in turn astronomically unlikely. As Fisher rightly notes, "the objection is more in the nature of an innuendo than of a criticism, for it depends for its force on the ambiguity of the word chance, in its popular uses" (Fisher, 1930, p. 37). The games in a casino, he notes, are of course "chancy," but there is nothing chancy about the fact that the house always wins in the long run. "It is easy without any very profound logical analysis to perceive the difference between a succession of favorable deviations from the laws of chance," like the gambler on a winning streak, "and, on the other hand, the continuous and cumulative action of those laws. It is on the latter that the principle of Natural Selection relies" (Fisher, 1930, p. 37). And the Fundamental Theorem lets us see how this is the case.

What, then, does the theorem actually say? As Fisher initially introduces it, it states that "the rate of increase in fitness of any organism at any time is equal to its genetic variance in fitness at that time" (Fisher, 1930, p. 35); his summary at the end of the chapter has "species" instead of "organism" (Fisher, 1930, p. 46). Unpacking this single sentence is not easy; it has been the subject of extensive debate in the historical, philosophical, and biological literature. Anya Plutynski is right, I think, to have argued that the theorem is strictly and indeed necessarily true, as long as we are careful about the meaning of both "fitness" and "variance" that Fisher has very precisely defined prior to introducing the Fundamental Theorem (Plutynski, 2006). Without entering too deeply into the technical detail, it is important to note that Fisher is not talking here about the actual average fitness value of a population: the mean fitness of a population can, of course, decrease, while the variance that Fisher equates with fitness in the Fundamental Theorem can only be positive or zero. This is not a contradiction, as some have argued. Fisher is, rather, expressing something like what are, for him, the "conditions on the possibility of evolutionary change via natural selection" (Plutynski, 2006, p. 61), in complete concordance with what we have already seen to be his position on selective change. Natural selection requires a store of mutations of individual small effect in order to operate—this is the "genetic variance in fitness" to which the Fundamental Theorem refers. And rather than population fitness as a whole, Fisher is referring to "a 'partial' change in mean fitness, or that part of fitness due to changes in the 'gene ratio' alone. [… Fisher is concerned with] changes in fitness that are proportional directly to changes due to the average effect of a gene substitution" (Plutynski, 2006, p. 74). The Fundamental Theorem, then, shows us exactly the action of natural selection as Fisher understood it. Selection of the small mutations within the population as expressed by the additive genetic variance would result in the increase of at least part of the population's overall fitness. Of course, other processes might impinge on overall fitness, too, resulting in a decrease in fitness in the overall population—and environmental change and the increasing adaptation of competitors will instantiate what we would now call a sort of red queen scenario where populations must constantly improve in order just to remain static (Fisher, 1930, pp. 41–42). But Fisher can now offer us a formalized theory, if one that would prove difficult to apply in any particular empirical case, that describes the way in which this kind of selection could—or better, must—act.

The next few chapters of *GTNS* reintroduce his theory of dominance and mutation as it had been developed in the other works we have already considered. He presents his modifier-based approach to the evolution of dominance. Contra the assertion of several later historians that his work instantiated the peak of "bean-bag genetics," where individual "genes for" particular characters are acquired and lost, he restates and reinforces here the idea that somatic changes, "although genetically determined, may be influenced or governed either by the environment in which the substitution is examined, or by the other elements in the genetic composition" (Fisher, 1930, p. 54). The data on Bateson's chickens make another appearance.

He next considers the structure of variation itself, again in terms already familiar to us. Mutations are "initially subject to the full force of random survival," but if they survive long enough in spite of the "fortuitous element in the survival and reproduction of their bearers," any and all "beneficial mutations have a finite probability, simply related to the benefit which they confer, of establishing themselves as permanent in the heredity of the species" (Fisher, 1930, p. 117). The formal apparatus underlying this view, however, is now significantly more sophisticated, as Fisher has begun considering the dynamics of the distribution of genes within a population as a way to reinforce his previous work. Yet again, this view of mutation confirms Fisher's opinion that neutral evolution and genetic drift are broadly unimportant, as "the very small range of selective intensity in which a factor may be regarded as effectively neutral suggests that such a condition must in general be extremely transient" (Fisher, 1930, p. 95). He then extends this same approach to gene distributions to consider various kinds of stable population equilibria under different selection pressures.

His treatment of specific examples of sexual selection and mimicry in the next several chapters needs not detain us here, but we should close this brief summary of *GTNS* by considering the last third of the book addressing eugenics. Fisher begins by arguing both that human beings are evolved creatures and that our mental and moral qualities are inherited biologically and refined by natural selection. But this poses an immediate problem—and for Fisher, it is a problem for the theory of natural selection itself, not a detour into social science or politics. Civilization, he argues, is clearly an advantage to its members, in its ability to specialize occupations and organize commerce, public order, and military intervention. Given that this is true, he continues, we would expect that once civilization had appeared, it would be massively selectively favored, run to fixation, and almost certainly prevail against any and all uncivilized neighbors. But this is, of course, not borne out by the historical record, insofar as we are not all currently vassals of the Babylonian or ancient-Egyptian empire—civilizations are in fact regularly overrun by their "non-civilized" competitors. What went wrong, then? Why do major civilizations continue to fail?

It is worth pausing here to realize that, far from being the first bit of a separate section of *GTNS* directed at humans, this challenge is for Fisher much more like those in the chapters on sexual selection and mimicry—we have here a perplexing phenomenon produced by a system subject to natural selection, one for which it will be difficult to offer a selective explanation, and hence we need to figure out a way either to resolve or dissolve the apparent tension. There is nothing different in kind, for Fisher, about mimicry in butterflies and the decline and fall of the Roman Empire.

Neither theories of social change developed by historians, nor a quick analogy with the social insects (the only possible non-human analogue, he writes, of human civilization) will suffice. Rather, he first spends a chapter asserting that fertility constitutes an important, heritable trait in humans, whether directly via biological characters or indirectly, mediated by the mental and moral factors that convince people to have children. In the next two chapters, then, he argues that advanced civilizations have worked to differentially select for lower fertility among their higher classes, insofar as success in higher social classes is often easier to acquire for those with fewer or no children. We should, therefore, he proposes in the brief final chapter, construct a system of social credits, whereby families are offered compensation proportional to their standard of living to encourage them to have more children, thus instituting a counter-selective pressure that could result in higher birth rates among the "better" members of society.

It is easy to see why Mazumdar would have pegged Fisher as a particularly odd sort of eugenicist. At least as presented in *GTNS*, the problems that he sees with the structure of contemporary civilizations are not ones of moral rot, lack of respect, family decay, or any of the other usual smears launched by eugenic "superiors" at the lower classes. It is rather a question of population structure, and seems to be of a piece with the other difficulties concerning selection that Fisher treats here. The strangeness of this argument was noticed by his colleagues. J. B. S. Haldane would write that "the social part is highly controversial," not least because "if you convince me I shall have to become an extreme form of socialist" (J.B.S.H. to R.A.F., April 29, 1930; Bennett, 1983, p. 209)—tongue planted firmly in cheek, as Haldane was at this point already a socialist, and would go on to be a vocal member of the Communist Party. Of course, this is not to say that more reactionary and scornful attitudes were not present elsewhere in Fisher's writings—he would write to Haldane the next year that "the practical point is to combat the idea that racial decay, or the differential birth-rate, or any other social phenomenon which we judge undesirable, is to be accepted fatalistically as the 'Will of Allah,' rather than tackled scientifically like *rabies*" (R.A.F. to J.B.S.H., May 1, 1931; Bennett, 1983, p. 214). Just as there is more to evolution than natural selection, there is more to eugenics than selection-driven population features.

Indeterminism, creativity, and physics in evolution: Fisher's philosophy of science

As I hope to have already demonstrated, Fisher has indeed developed a theory of natural selection that emphasizes exactly those biometrical desiderata that I laid out over the last few chapters. But the way in which he does so is—as was the case with each of the authors who we have already discussed—quite philosophically sophisticated, and I want to end this chapter by examining what kinds of new philosophical moves Fisher made that, in turn, created room for his theoretical work. Four such threads, I think, are of particular importance. First, Fisher engages in a remarkable project of revisionist history, on which most major discoveries of the preceding several decades are recast as the result of formal statistical methodology like his own. Second, Fisher's introduction of statistical physics serves as a key element in his pursuit of a certain kind of theoretical abstraction, interestingly different from that which we found in the case of the biometrical school. Third, Fisher is the first to clearly argue that the probabilistic and statistical conclusions to be drawn in the case of natural selection were often conclusions about hypothetical populations, not actual ones, and to work out what this meant for the relationship between such theorizing and empirical observation. Fourth, and finally, we must return to the question of Fisher's indeterministic view of causation.

Revisionist history

To begin, and to better situate Fisher within the structure of this book so far, it bears mention that he engages in some fairly impressive pieces of revisionist historical work in order to cast a statistical approach to biology as a central element of the recent history of evolutionary science. First we have the case of Mendel. "It deserves notice," Fisher writes, "that the first decisive experiments [in genetics], which opened out in biology this field of exact study, were due to a young mathematician, Gregor Mendel" (Fisher, 1930, p. viii). Placing Mendel as a statistical pioneer was a longstanding goal of Fisher's; as early as 1924 he had written in a review of biometry for a popular audience that "Mendel was in fact the pioneer of the statistical method in the study of heredity; and it happened that the statistical method of research was the fruitful one" (Fisher, 1924, p. 190). Of course, it is true that Mendel experimented upon a large number of replicates, counted up the results of his experiments and compared their ratios, developed a symbolic system for keeping track of those results, and that this turns out to be a very important way to approach the outcomes of breeding experiments. But interpreting Mendel as a "mathematician" places him, following on a number of recent historical investigations, in the wrong tradition—his work was not much out of the ordinary for contemporary studies in plant breeding and the analysis of hybridization, a particularly strong tradition in German-speaking countries of the day (see, e.g., Müller-Wille and Rheinberger, 2012, pp. 129–132).

Perhaps even more impressively, Fisher performed the same retroactive trick with the discovery of chromosomes. By arguing that the crucial experiment for the chromosomal theory of heredity was the discovery of linkage, he could then claim that such a discovery "requires a considerable number of offspring to detect it, and still more to make an accurate estimate of the percentage of crossovers," leading chromosomes as well to be "the product of the statistical method" (Fisher, 1924, p. 190).

One might be tempted to accuse Fisher of biological naiveté here, certainly a popular complaint about Fisher's work. Ignoring the importance of cytological, developmental, and non-statistical theoretical work to the discovery of chromosomes might then be read as yet another way in which Fisher looks to reduce all of the interesting phenomena in the life sciences to statistical computation. (This critique would have sat particularly well with Punnett.) But I think this would be short-sighted. The rewritten history of recent evolutionary theorizing serves a number of important goals for Fisher, and as we have seen above, it is not as though he is simply unaware of the biological underpinnings of his theoretical work. We should thus take seriously the idea that this is a deliberate historiographical and philosophical move on Fisher's part. First, it makes the kind of work that he wanted to pursue nearly inevitable. Everyone wants to be on the right side of history. But second, and more importantly, it allows him to separate himself cleanly from the polemical partisans he found on both sides—Pearson, Bateson, and Punnett. It is this sort of distance that lets him dismiss the Pearson-Bateson argument, in which he will not be mired, as "one of the most needless of the controversies in the history of science" (Fisher, 1924, p. 189). While he does so in a relatively heavy-handed manner, we can profitably read Fisher as engaging in a kind of explicit community and discipline formation of the sort that historians of science have, time and again, noted was so important to the Modern Synthesis (Cain, 1994; Smocovitis, 1996).

Abstraction and statistical physics

Another striking feature of Fisher's program in *GTNS* is its self-conscious focus on abstraction and generality. His aim, even more directly than was the intent of Weldon or the arch-positivist Pearson, is to build a conception of selection that holds in any population whatsoever, unencumbered by the contingency of the biological world. Interestingly, he approaches this in explicitly modal terms:

> *The ordinary mathematical procedure in dealing with any actual problem is, after abstracting what are believed to be the essential elements of the problem, to consider it as one of a system of possibilities infinitely wider than the actual, the essential relations of which may be apprehended by generalized reasoning, and subsumed in general formulae, which may be applied at will to any particular case considered.*

(Fisher, 1930, p. ix)

Our goal in constructing a theory of natural selection is to describe the space of possibilities in which evolution operates, and then to express the processes operating within that space, like natural selection, as relations (here, statistical relations) between those possibilities. Having done so, we will be able to apply the resulting theoretical framework to whatever natural situations might later present themselves.

We can see here a number of threads already teased apart above. Most importantly, it gives us a clear understanding of why it is that the reconciliation between biometry and Mendelism was the theoretical foundation which Fisher took to be necessary for the construction of the Fundamental Theorem. Far from being two separate projects, one of solving the question of inheritance and one of solving the question of evolutionary change, for Fisher, as for many of the other authors we've considered, the two are intimately and necessarily connected. The picture of variation as, in general, small effects brought about in systems containing large numbers of Mendelian factors responsible for the production of phenotypic traits is exactly what we need to establish the possibility space in which natural selection works. This is made particularly clear by his later considerations of stable equilibria under selection pressures, worked out precisely in terms of distributions describing the number of mutant genes present within a particular population—we are here looking at evolutionary dynamics as privileging certain classes of these possibilities as they were drawn out in Fisher's underlying synthesis of biometrical and Mendelian understandings of inheritance. And the Fundamental Theorem, then—recalling Plutynski's invocation of the "conditions on the possibility of natural selection"—appears as a way to pick out certain patterns within the changes in those underlying distributions, or in Fisher's words, the way in which we apprehend them by generalized reasoning and subsume them in general formulae.

It is here that Fisher's grounding in statistical physics becomes most relevant. Inspired by the survey of his uses of physics developed by Hodge (1992), I want to divide them into three different types, two of which are particularly pertinent to Fisher's pursuit of theoretical generality.

Hodge begins by unpacking a 1915 article that Fisher wrote with C. S. Stock, which makes frequent and illustrative use of a variety of metaphors connecting evolution with statistical mechanics (Fisher and Stock, 1915). First, evolution and thermodynamics are alike in that it is a statistical property which guarantees the stability of their results in the first place: "the reliability and predictability of the outcome," that is, emerges "when the individuals are numerous and the causes acting upon them independent" (Hodge, 1992, p. 248). The same kind of stability is invoked in the summary of the 1922 paper on selection, where Fisher writes that "'the distribution of the frequency ratio' for different hereditary factors is—in the absence of selection and random survival effects and so on—a stable one like that of 'velocities in the Theory of Gases'" (Hodge, 1992, p. 249). In both cases, it is the very fact that evolving systems form a statistical population that lets us draw certain kinds of conclusions from them, or, again, apprehend their relations by generalized reasoning.

A second form of generality can also be guaranteed by analogy with statistical physics—in mechanics, we are "independent of particular knowledge about separate atoms, as in eugenics we are independent of particular knowledge about individuals" (Fisher and Stock, 1915, pp. 60–61; see Hodge, 1992, p. 248). In the same way, when Fisher wrote the 1918 paper, he thought that "assumptions about the dominance of particular factors, about the size of their effects, about their proportion in the population, about dimorphism and polymorphism, and about linkage" (Hodge, 1992, pp. 248–249) made general theory construction impossible, and so, he would write later, what was needed was an investigation not unlike "the analytical treatment of the Theory of Gases, in which it is possible to make the most varied assumptions as to the accidental circumstances, and even the essential nature of the individual molecules, and yet to develop the general laws as to the behavior of gases, leaving but a few fundamental constants to be determined by experiment" (Fisher, 1922a, pp. 321–322). Indeed, it is this obsessive pursuit of generality, I think, that leads Fisher to often be loose with just what it is that we are abstracting away from, be it genes, individuals, populations, or species—recall that, as was noted in passing above, Fisher's arguments surrounding the Fundamental Theorem "repeatedly fail to make clear whether the 'organism' Fisher is discussing is a single individual or a whole population or even species" (Turner, 1987, p. 327). We can of course reconstruct after the fact what Fisher must have had in mind to make his mathematics consistent, but his drive to abstraction in the manner of statistical mechanics seems to make him a little too quick to neglect the units which underlie statistical change.

Third, and I think least interestingly, we have an analogy that Fisher draws in *GTNS* during his explication of blending inheritance. "The particulate theory of inheritance," he writes, "resembles the kinetic theory of gases with its perfectly elastic collisions, whereas the blending theory resembles a theory of gases with inelastic collisions, and in which some outside agency is required to be continually at work to keep the particles astir" (Fisher, 1930, p. 11). Hodge goes on to connect this to the Fundamental Theorem, arguing that in the way that "the Second Law [of thermodynamics] concerns the progressive loss of energy available for work," the Fundamental Theorem similarly "relates the rate of increase in fitness…to the genetic variance in fitness that selection consumes in producing that effect" (Hodge, 1992, pp. 250–251). I worry, however, that this is to read too much into an analogy to which Fisher does not refer again for the rest of *GTNS*. The distinction between the mechanisms of particulate and blending inheritance is, Fisher writes, like the distinction between inelastic and elastic collisions only insofar as elastic collisions constitute a process which, like the reduction of variance as a result of blending inheritance, saps the system of all its variability and tends to quickly reduce it to a static end state.

The real utility of analogies with statistical physics, from the second law to the stability and reliability of the observed outcomes, lies in their capacity to reinforce Fisher's pursuit of theoretical generality.

Unfortunately for this clean picture, Fisher complicates this understanding later on in *GTNS*, when he turns to the example of sexual selection. How might we come to know, he wonders, which process was actually responsible for some given instance of evolutionary change? I will turn to the question of the empirical evidence that we would require in the next section. But as for the theoretical work that might allow us to successfully make such a distinction, Fisher urges caution. As he later critiques supposed statistical limitations to his understanding of mimicry, he writes that

> *An evolutionary tendency is perceived intuitively, and expressed in terms which simplify, and therefore necessarily falsify, the actual biological facts. The only reality which stands behind such abstract theories is, in each case, the aggregate of all the incidents of a particular kind, which can occur from moment to moment to members of a species in the course of their life-histories.*
>
> **(Fisher, 1930, p. 150)**

It would be easy to read this as a sort of Pearsonian anti-realism, a commitment to the idea that evolutionary trends, insofar as they are abstracted away from their details by individual scientists with particular purposes in mind, have no reality in and of themselves in the biological world, and are supported only by particular cases of living and dying. But in the first of the chapters on the decline of human societies, while evaluating the unsuitability of prior "sociological" theories of human cultural change, Fisher puts the point differently:

> *Generalized description should, however, never be regarded as an aim in itself. It is at best a means towards apprehending the causal processes which have given rise to the phenomena observed. Beyond a certain point it can only be pursued at the cost of omitting or ignoring real discrepancies of detail, which, if the causes were understood, might be details of great consequence.*
>
> **(Fisher, 1930, p. 178)**

I will return to this talk of causation further on. But we now seem to have Fisher pulling us in three different directions. Straightforward readings of these three long passages seem to have him arguing for a sort of modal approach to scientific theorizing as describing processes operating within larger spaces of possibility; for a kind of anti-realist, perspectival view of evolutionary theorizing as interest-relative and non-causal; and for general theory as one tool among many, useful in our genuinely fundamental search for the true causal processes operating in nature. Which is it?

I think there remains a way to square the circle here, taking insight from the kind of detail-inclusive statistical causation that, as I argued in Chapter 4, we also find in the work of Weldon. The first of the three Fisher passages above, concerning the modal structure of scientific theorizing, can be interpreted fairly literally. There, we once again see Fisher's debt to theoretical physics—describing evolutionary systems in terms of state spaces, and the processes operating on them in terms of various ways of subdividing change within those state spaces. The key to the second quotation, then, is to realize what Fisher means when he speaks of an "evolutionary tendency." This arises in the midst of his discussion of mimicry, a longstanding debate and precisely the sort of case where evolutionary change is described not in a clear, formally sound, statistical manner, but rather as a sort of rough intuition: we see that a number of species of a certain genus of butterfly appear to have evolved identical coloration, and we can make the rough guess that delicious mimics have therefore profited from the unpalatability of their look-alikes. How are we to move beyond such vague descriptions? We have to come to understand the aggregate of events which have actually impinged upon those organisms (in this case, which might let us distinguish the various ways in which mimicry has evolved), and that is precisely what we are supposed to do with the aid of the statistical theory that Fisher has devised.

The third quotation, then, is Weldonian through-and-through. We are not generalizing for the sake of generalizing—we are generalizing because it is precisely here that we can hope to have some degree of cognitive access to processes of the mass-action, population type of which natural selection is taken to be a paradigmatic example. As Fisher would write in discussing the reticence of Darwin and Wallace to publish their evolutionary theorizing prior to uncovering the mechanism of natural selection, "once the nexus of detailed causation was established, evolution became not merely History, but Science" (Fisher, 1930, p. 179). Of course, we must remain constantly on our guard. We may find that some of the detail that we have abstracted away from in fact hides an important influence on future population change. But—to reinforce the point made above about the proper interpretation of the Fundamental Theorem—Fisher always recognized this. The reason that the Fundamental Theorem excited him was not that it would provide the sole explanation for evolutionary change, but rather that it would be precisely the kind of tool he needed to separate a particular kind of selective change from the myriad other events that govern the dynamics of evolving populations.

Hypothetical populations and empirical evidence

For Weldon, however, as we saw at the end of Chapter 4, the distributions upon which this theoretical abstraction was to work its magic were always distributions of actual organisms, in real-world circumstances—distributions of characters like the frontal breadth of his crabs. Fisher is the first to make a crucial move, occasionally hinted at by Galton and others but never before made rigorous, to theorizing about hypothetical populations, taking seriously the idea that evolutionary theory is to be defined over the space of possible organisms and gene combinations and only later instantiated in particular cases.

As he begins to introduce the groundwork for the Fundamental Theorem, he notes that estimated values of fitness may in many cases be inaccurate and difficult to obtain. "As in all other experimental determinations of theoretical values," he writes, "the accuracy attainable in practice is limited by the extent of the observations; the result derived from any finite number of observations will be liable to an error of random sampling" (Fisher, 1930, p. 23). There will thus be practical, epistemic limits on the particular numerical quantities we assess for life-history parameters. But this shouldn't cause us any undue worry. "This fact does not, in any degree," he continues, "render such concepts as death rates or expectations of life obscure or inexact. These are statements of probabilities, averages &c., pertaining to the hypothetical population sampled, and depend only upon its nature and circumstances" (Fisher, 1930, p. 23). For the first time here, then, Fisher is using a distinction which he has drawn in the analysis of statistical data to clarify the nature of evolutionary theorizing. It was Fisher who had first separated, in the context of statistical analysis in 1922, "the *hypothetical infinite population* whose *parameters* we should like to know and…the *sample* whose *statistics* are our estimates of the parameter values" (Box, 1978, p. 90; discussing Fisher, 1922b). The immediate consequence for a statistical theory of evolution is that it is this hypothetical population that bears fitness values or selection coefficients. While we can only offer estimates of those real values, that does not cast doubt on their underlying reality. The quantities that would result from considering "not merely the whole of a species in any one generation attaining maturity, but…all the genetic combinations possible, with frequencies appropriate to their actual probabilities of occurrence and survival" will therefore yield an "exact" concept, "not dependent upon chance as must be any practical estimate of it" (Fisher, 1930, pp. 30–31).

This opens up a particular challenge for Fisher, however, hinted at in this last quote: if one takes the target of one's theorizing to be the hypothetical population rather than the actual one, we are owed an account of just how it is that empirical data can help us come to understand those hypothetical populations. Is there anything more sophisticated to be said than that we should attempt to obtain the most accurate estimates we have available to us of the underlying theoretical values? It turns out, for Fisher, that there is—or, at least, that we should approach the question with caution. As Fisher considers how we might disentangle the effects of natural selection from those of sexual selection acting on a given population, he writes that, by comparison with the kinds of population statistics required in the demonstration of the Fundamental Theorem,

> The distinction between one kind of selection and another would seem to require information in one respect infinitely more detailed, for we should require to know not the gross rates of death and reproduction only, but the nature and frequency of all the bionomic situations in which these events occur.
>
> **(Fisher, 1930, p. 132)**

While he never precisely defines "bionomic situations" in *GTNS*, it seems as though he intends them to be the ingredients that make up an environment—the precise circumstances of organism-environment interaction in which a particular individual might find itself at a particular time, specified enough that their outcomes could plausibly be said to be governed by natural law (hence "bionomic"). An organism's environment, then, is just the collection of bionomic situations to which it might be subject in a given range of times and places. Fisher is therefore worried that a theory which could separate all the various processes that might have an effect on population dynamics would require an amount of knowledge that would be impossible to obtain about any natural population. "In the vast majority of cases," he writes, "the evidence will be too scanty to be decisive" (Fisher, 1930, p. 132).

Of course, we are not left with only despair—in particular cases, features of those situations might be generalizable in such a way as to allow for the determination of the relative influence of various processes. "I should not like to be taken," he demurs, "to be throwing doubt on the value of such distinctions as can be made" (Fisher, 1930, p. 132). But it is clear that Fisher is concerned that in fact the framework he has built will not give us enough leverage to make the kinds of attributions of evolutionary change to particular causes that he would have liked in real-world scenarios.

This is, then, Fisher's quest for general theory pushed to the extremes—perhaps even too far for Fisher himself. Having hoped to derive a model of evolutionary change that was applicable to any possible natural population, such a theory deals perforce in state spaces, possibilities, and hypothetical populations, and will therefore only with difficulty be applicable to the kind of fine-grained analysis of causal influence so often important in evolutionary investigations. One is reminded of Richard Levins's classic work on the trade-offs implicit in biological modeling, with Fisher having clearly chosen to "sacrifice realism to generality and precision" (Levins, 1966, p. 422).

Causation and indeterminism

The last, and certainly most interesting, element of Fisher's philosophy of science that is worth drawing out here is his deep and abiding commitment to indeterminism. For Hodge, indeed, more than perhaps any other feature of his philosophical or scientific outlook, it is indeterminism that can tie together Fisher's views into a coherent and comprehensible package (Hodge, 1992, pp. 256–259). Conveniently for us, Fisher published a paper detailing exactly his position on indeterminism and causation in the first-ever volume of the journal *Philosophy of Science*, just a few years after the appearance of *GTNS* (Fisher, 1934).

Fisher's understanding of indeterminism cuts across several categories that we might separate in a contemporary context, and accomplishes for him a variety of different goals; hence it merits careful unpacking. First, there is a straightforward contrast to be drawn between what we might call deterministic and indeterministic approaches to causation. As Fisher puts it, in the deterministic, Newtonian universe, "the future state of a system could be calculated rigorously, supposing its initial state were known with absolute precision" (Fisher, 1934, p. 101). The results of these calculations so precisely fit future observations that it was quickly assumed that such formalism must have simply uncovered the true laws of nature:

> The mathematical formulae were constantly found to be more accurate than the observations upon which they had first been based, or, when their form had to be modified, they were found to be so much the more comprehensive. It is not to be wondered at that experimenters and theorists alike should have regarded it as axiomatic that *there* were *laws of causation*, similar in form to those already established, which controlled the happenings of the world with perfect rigor.
>
> **(Fisher, 1934, p. 102)**

Of course, the apple-cart was upset by the advent of successful statistical methodologies in physics at the end of the 19th century. Such theories resisted any framing in deterministic terms, and provided us probabilistic predictions that were every bit as reliable as those in deterministic sciences. Indeterminism was on the march.

This might, of course, have been a point of concern: can we really live in an indeterministic world, much less do science in one? Here again, Fisher owes a sort of debt to the thermodynamics he learned from Jeans. "The great 19th-century developments in kinetic theory show," as Hodge puts it, "that indeterminism is entirely consistent with the orderliness in the natural and social worlds and with success in the quest for knowledge, including causal knowledge, of those worlds" (Hodge, 1992, p. 258). We see this in two different ways. First, Fisher writes, our everyday experience of causation is itself indeterministic, and hence indeterminism is a view "we have always known to be true in our daily affairs" (Fisher, 1934, p. 104). Effectively no human experience arises as a matter of law; we ought not, Fisher implies, have expected science to be any different. Second, and more importantly—again the quest for theoretical generality!—we should see that determinism is only a special case of the probabilistic influence of prior events on latter events, and therefore indeterministic causation, or "the view that prediction of [the] future from past observation must always involve uncertainty, and, when stated correctly, must always be a statement of probabilities, has the scientific advantage of being a more general theory of natural causation" (Fisher, 1934, p. 104). We can therefore take indeterminism on board without having to claim that deterministic prediction is never applicable; we find ourselves with a strict increase in theoretical power.

Many potential effects of such a perspective might result—not least among them, he writes, the possibility that biologists will be less envious of the "exact" sciences (Fisher, 1934, p. 104)—but we must be sure to underline that indeterminism

> does not in the least imply an anarchy of causelessness...natural law is none the less real if, when precisely stated, it turns out to be a statement of probability: causation is none the less recognizable, and an action is just as much an effective cause of subsequent events, if it influences their respective probabilities, as if it predetermines some one of them to the exclusion of the others.
>
> **(Fisher, 1934, pp. 105–106)**

So much for the implications of indeterministic causation. Between the already significant use of probabilistic causes in the physical sciences and the fact that the adoption of indeterminism in this sense makes no trouble at all for extant deterministic explanations or the notion of causation in general, Fisher sees no reason that we ought not to be willing to generalize our philosophy of science in this way.

But this is not all that Fisher means by indeterminism. For the fact that the future is only probabilistically constrained by the past offers Fisher a place to ground free will and active choice, not only for humans, but in the biological realm as well. On an indeterministic Darwinism, "creative causation is centered in the organisms themselves," which must produce "choice and spontaneous action" that is selectively valuable "in its harmony with the world around, its capacity to utilize its advantages, or penetrate its undiscovered possibilities" (Fisher, 1934, p. 111). The genius of Darwin's theory of selection lies in "locating the driving force of the evolutionary process in this manifold contact between the inner and the outer worlds" (Fisher, 1934, p. 111). Organisms—humans included—are thus implicated on an indeterministic world-view as active agents with a part to play in the construction of their own evolutionary futures.

It is here, as a number of authors have persuasively argued, that Fisher can reconcile his thermodynamic and his biologico-eugenical commitments. As Hodge puts it, Fisher's world is governed by two processes, each represented by a scientific hero: "Boltzmann is right about the way down; entropic decline rules everywhere. The only exception is the living world which thanks to natural selection (the only counterentropic cause in the universe) is always on the way up" (Hodge, 2011, p. 36; see also Turner, 1985). Thermodynamic decay is to be offset by selective freedom and construction, and it is Fisher's indeterminism that allows these two processes to be welded together in a coherent whole.

While we can certainly claim that Fisher's approach to natural selection has finally succeeded at fulfilling the aims of the biometricians as we saw them unfold over the last several chapters, it is notable that he has added to their picture several further ingredients that are, I think, indispensable for its eventual success. Fisher's hypothetical-modal approach to evolutionary theorizing allows him to craft an understanding of natural selection at a sufficient level of generality to avoid the pitfalls that can be brought on by an over-reliance on particular empirical case studies (think, for instance, of Weldon's having to normalize for the growth rate of his crabs in order to know whether distribution change was genuinely due to natural selection). Fisher's embrace of indeterminism and his fluency in contemporary statistical physics removed the philosophical barriers which might otherwise have prevented him from being willing to treat a formula as esoteric as the Fundamental Theorem as a candidate for being a law of nature. And, in the end, the power and clarity of his theoretical apparatus would bring statistical theorizing to evolutionary theory for good. To the chagrin of Bateson's ghost, the pompous parade of arithmetic rolls on.

References

Bennett, J.H. (Ed.), 1983. Natural Selection, Heredity, and Eugenics: Including Selected Correspondence of R.A. Fisher with Leonard Darwin and Others. Clarendon Press, Oxford.

Bowler, P.J., 2009. Science for All: The Popularization of Science in Early Twentieth-Century Britain. University of Chicago Press, Chicago.

Box, J.F., Fisher, R.A., 1978. The Life of a Scientist. John Wiley & Sons, New York.

Brownlee, J., 1910. The significance of the correlation coefficient when applied to Mendelian distributions. Proc. R. Soc. Edinb. 30, 473–507. https://doi.org/10.1017/S0370164600030935.

Cain, J., 1994. Ernst Mayr as community architect: launching the Society for the Study of Evolution and the journal Evolution. Biol. Philos. 9, 387–427. https://doi.org/10.1007/BF00857945.

Depew, D.J., Weber, B.H., 1995. Darwinism Evolving: Systems Dynamics and the Genealogy of Natural Selection. Bradford Books, Cambridge, MA.

Dronamraju, K., 2017. Popularizing Science: The Life and Work of JBS Haldane. Oxford University Press, Oxford.

Edwards, A.W.F., 2013. Robert Heath Lock and his textbook of genetics, 1906. Genetics 194, 529–537. https://doi.org/10.1534/genetics.113.151266.

Fisher, R.A., 1947. The renaissance of Darwinism. Listener 37, 1001.

Fisher, R.A., 1937. Professor Karl Pearson and the method of moments. Ann. Eugen. 7, 303–318. https://doi.org/10.1111/j.1469-1809.1937.tb02149.x.

Fisher, R.A., 1934. Indeterminism and natural selection. Philos. Sci. 1, 99–117.

Fisher, R.A., 1929. The evolution of dominance; reply to professor Sewall Wright. Am. Nat. 63, 553–556. https://doi.org/10.1086/280289.

Fisher, R.A., 1930. The Genetical Theory of Natural Selection. Clarendon Press, Oxford.

Fisher, R.A., 1928a. The possible modification of the response of the wild type to recurrent mutations. Am. Nat. 62, 115–126. https://doi.org/10.1086/280193.

Fisher, R.A., 1928b. Two further notes on the origin of dominance. Am. Nat. 62, 571–574. https://doi.org/10.1086/280234.

Fisher, R.A., 1927. On some objections to mimicry theory; statistical and genetic. Trans. R. Entomol. Soc. Lond. 75, 269–278. https://doi.org/10.1111/j.1365-2311.1927.tb00074.x.

Fisher, R.A., 1924. The biometrical study of heredity. Eugen. Rev. 16, 189–210.

Fisher, R.A., 1922a. On the dominance ratio. Proc. R. Soc. Edinb. 42, 321–341. https://doi.org/10.1016/S0092-8240(05)80012-6.

Fisher, R.A., 1922b. On the mathematical foundations of theoretical statistics. Philos. Trans. R. Soc. Lond. A 222, 309–368. https://doi.org/10.1098/rsta.1922.0009.

Fisher, R.A., 1918. The correlation between relatives on the supposition of Mendelian inheritance. Philos. Trans. R. Soc. Edinb. 52, 399–433.

Fisher, R.A., Stock, C.S., 1915. Cuénot on preadaptation: a criticism. Eugen. Rev. 7, 46–61.

Hodge, M.J.S., 2011. Darwinism after Mendelism: the case of Sewall Wright's intellectual synthesis in his shifting balance theory of evolution (1931). Stud. Hist. Philos. Biol. Biomed. Sci. 42, 30–39. https://doi.org/10.1016/j.shpsc.2010.11.008.

Hodge, M.J.S., 1992. Biology and philosophy (including ideology): a study of Fisher and Wright. In: Sarkar, S. (Ed.), The Founders of Evolutionary Genetics. Kluwer Academic Publishers, Dordrecht, pp. 231–293.

Levins, R., 1966. The strategy of model building in population biology. Am. Sci. 54, 421–431.

MacKenzie, D.A., 1981. Statistics in Britain, 1865–1930: The Social Construction of Scientific Knowledge. Edinburgh University Press, Edinburgh.

Mazumdar, P.M.H., 1992. Ideology and method: R. A. Fisher and research in eugenics. In: Eugenics, Human Genetics, and Human Failings: The Eugenics Society, Its Sources and Its Critics in Britain. Routledge, London, pp. 96–145.

Moore, J., 2007. R. A. Fisher: a faith fit for eugenics. Stud. Hist. Philos. Biol. Biomed. Sci. 38, 110–135. https://doi.org/10.1016/j.shpsc.2006.12.007.

Moran, P.A.P., Smith, C.A.B., 1966. Commentary on R. A. Fisher's paper on the correlation between relatives on the supposition of Mendelian inheritance. In: Eugenics Laboratory Memoirs. Cambridge University Press, Cambridge.

Morrell, J., 1997. Science at Oxford, 1914–1939. Oxford University Press, Oxford.

Müller-Wille, S., Rheinberger, H.-J., 2012. A Cultural History of Heredity. University of Chicago Press, Chicago.

Parolini, G., 2015a. The emergence of modern statistics in agricultural science: analysis of variance, experimental design and the reshaping of research at Rothamsted Experimental Station, 1919–1933. J. Hist. Biol. 48, 301–335. https://doi.org/10.1007/s10739-014-9394-z.

Parolini, G., 2015b. In pursuit of a science of agriculture: the role of statistics in field experiments. Hist. Philos. Life Sci. 37, 261–281. https://doi.org/10.1007/s40656-015-0075-9.

Pearson, E.S., 1938. Karl Pearson: an appreciation of some aspects of his life and work. Part II. 1906–1936. Biometrika 29, 161–248. https://doi.org/10.1093/biomet/29.3-4.161.

Pearson, K., 1904. Mathematical contributions to the theory of evolution. XII. On a generalized theory of alternative inheritance, with special reference to Mendel's laws. Philos. Trans. R. Soc. Lond. A 203, 53–86. https://doi.org/10.1098/rsta.1904.0015.

Plutynski, A., 2006. What was Fisher's fundamental theorem of natural selection and what was it for? Stud. Hist. Philos. Biol. Biomed. Sci. 37, 59–82. https://doi.org/10.1016/j.shpsc.2005.12.004.

Punnett, R.C., 1930. [Review of] The genetical theory of natural selection. Nature 126, 595–597. https://doi.org/10.1038/126595a0.

Punnett, R.C., 1915. Mimicry in Butterflies. Cambridge University Press, Cambridge.

Punnett, R.C., 1907. Mendelism, second ed. Bowes and Bowes, Cambridge.

Smocovitis, V.B., 1996. Unifying Biology: The Evolutionary Synthesis and Evolutionary Biology. Princeton University Press, Princeton, NJ.

Snow, E.C., 1910. On the determination of the chief correlations between collaterals in the case of a simple Mendelian population mating at random. Proc. R. Soc. Lond. B 83, 37–55.

Turner, J.R.G., 1987. Random genetic drift, R. A. Fisher, and the Oxford school of ecological genetics. In: Krüger, L., Gigerenzer, G., Morgan, M.S. (Eds.), The Probabilistic Revolution, Volume 2: Ideas in the Sciences. Bradford Books, Cambridge, MA, pp. 313–354.

Turner, J.R.G., 1985. Fisher's evolutionary faith and the challenge of mimicry. Oxf. Stud. Evolut. Biol. 2, 159–219.

Wright, S., 1929. Fisher's theory of dominance. Am. Nat. 63, 274–279.

Yule, G.U., 1906. On the theory of inheritance of quantitative compound characters on the basis of Mendel's laws—a preliminary note. In: Wilks, W. (Ed.), Report of the Third International Conference on Genetics. Spottiswoode & Co, London, pp. 140–142.

Conclusions, historiographical and philosophical

7

There is more, far more, to the history of statistical thinking than the history of mathematical statistics.
Jon Hodge, 1992

It is here that I will end my story. My goal has been to describe a broad conceptual shift which re-wired our understanding of the living world: the change in evolutionary theory that took place over the century from 1830 to 1930, from a version of evolution that could be presented without any real reference to chance or any statistical formalism whatsoever, to our now-contemporary view on which evolution cannot even be envisaged without the methods of statistics and the concepts of chance and probability that underlie their use. As I argued in the introduction, there are myriad different ways in which we might approach the transmission of those ideas from their birth in the 19th century to today's theory of evolution, or more specifically, of connecting the works that I discussed in the first four chapters with currents in the Modern Synthesis. Here, I have drawn that link through the lens of R. A. Fisher. The publication of Fisher's *Genetical Theory of Natural Selection* marks, as cleanly as one might hope, a point at which these methods and the philosophy of science which justified them can be said to have entered into the mainstream. Fisher's work would become a part of a newly formed canon of evolutionary thought, along with the contributions of those like Wright, Haldane, Dobzhansky, Mayr, Simpson, and Stebbins (Plutynski, 2009), which through an exceptionally conscious and explicit process of discipline formation, society building, journal founding, and so on (see, especially, the masterful works of Smocovitis, 1994a,b, 1996) became the new field of evolutionary biology, forming what those actors themselves began to call the Modern Synthesis.

An assessment of the role of chance, probability, and statistics in the context of the broader, technically sophisticated, and complex landscape of the Modern Synthesis—or in the context of the development of molecular biology and biochemistry which would shortly follow it (Monod, 1971; Sloan and Fogel, 2011)—would require at least another work of this length, and a scholar of that period of better quality than myself (for an excellent summary of the major issues, see Plutynski et al., 2016). In this final chapter, I want to instead do three things. First, I will very briefly summarize the book's major philosophical and historiographical themes. Then, I

The Rise of Chance in Evolutionary Theory. https://doi.org/10.1016/B978-0-323-91291-4.00001-7

compare the approach that I have taken with two other works that have offered similar narratives and which I have not yet had reason to discuss in detail—Jean Gayon's *Darwinism's Struggle for Survival* and David Depew and Bruce Weber's *Darwinism Evolving*. Finally, I return to my own position, sketching what I take to be some broad trends that have surfaced, time and again, throughout the book; I hope they will not be old news to the attentive reader. In doing so, I will also offer some thoughts on future work and on other approaches to this question that would be complementary to my own. As Gayon aptly put it, mine "is not *the* history, but rather *a* history" (Gayon, 1998, p. 15) of this important conceptual shift.

A quick look back

Perhaps the most significantly unorthodox commitment of my work here is that to a broadly continuous historical view of this period, commonly marked as one of debate and discord, if not outright minor revolution in the Kuhnian sense. One aspect of this view is rooted in my skepticism about the standard framing of the years from 1880 to 1910 as the "biometry-Mendelism debate," a point to which I will return in the last section of this chapter. I join with a small but growing chorus of historians of biology in arguing that this perspective is reductive, overly constraining, and generally past its sell-by date.

But what should replace it? I have here pushed for another non-standard view: that the insights of biometry remained, if not central to evolutionary biology across this period, at the very least familiar to all of its practitioners, in a way that is not always recognized by historians and philosophers of biology. That, in turn, means that not only the technical tools of mathematical statistics which they developed but also, I argue, their work building a philosophical approach on which the use of those technical tools could be justified, would be carried over into the Modern Synthesis. Even an author who offers an interpretation of evolution as sophisticated and innovative as that of R. A. Fisher does so in a context deeply informed by the picture of natural selection developed over the preceding years.

Finally, I have very explicitly endeavored here *not* to discuss another of the most common ways in which this period is presented. Namely, I think the story that I have told here is one that can stand on its own, in the absence of a detailed analysis of the concepts of *heredity* that were so important to the study of evolution during this time. Of course, the nature of heredity is an unavoidable problem, and I have—especially in considering the work of Galton or Weldon's final research project—by no means avoided the question entirely in my presentation. But I think there is room for histories of this era that do not make heredity their primary axis of analysis, and I hope to have demonstrated that by example. In the same way that, for some authors, heredity is the fundamental category and positions on natural selection follow after, I think the narrative here, on which questions of chance and their relationship to selection are primary, and views on heredity follow after, is an equally coherent one, and one which illuminates hitherto unappreciated aspects of the biological and philosophical work of the figures I've considered.

This is only a brief summary of a few of the most unusual features of my story; I will return to them in greater detail in the final section of this chapter. But it will suffice for what I think is now a very important undertaking: a comparison of my work with two other volumes that can be said to have offered synthetic approaches to the same developments that I have highlighted here.

A comparative interlude: Gayon & Depew and Weber

Two central commitments structured Jean Gayon's *Darwinism's Struggle for Survival* (1998). The first such position should be apparent from the title, revised from the original French edition's less combative *Darwin et l'après-Darwin* (Gayon, 1992). "The modern theory of natural selection," he writes, "arose out of a long initial crisis, the first signs of which can be discerned in Darwin's writings, and which had barely been overcome by the 1920s" (Gayon, 1998, p. 5). This "eclipse of Darwinism"— Gayon freely borrows the phrase, coined by Julian Huxley (1942) and expanded and popularized by the wide-ranging study of "anti-Darwinian" approaches to evolution by Peter Bowler (1992)—constitutes one of the main problems to be solved, Gayon argues, for historians and philosophers of biology. Natural selection, as we present it in today's textbooks, seems to be intuitively plausible, widely demonstrated, and well understood. Why, then, did it face so much opposition and competition for so long? For Gayon, responding to this cluster of questions must be one of the central goals of any history of this period.

Thus phrased, however, this problem still leaves a number of open interpretive points. To say that "Darwinism" was eclipsed requires that we have a solid definition of just what "Darwinism" would have meant across this long and rapidly changing period. As we have seen, such a definition may emerge only with difficulty: recall from Chapter 2 that Galton's treatment of natural selection was so muddled that even his contemporaries had trouble discerning whether he actually supported a gradualist or a saltationist perspective (Bowler, 2014). Gayon, to be sure, recognizes how problematic this concept is (he notes, in passing, that matters only become worse if one expands one's scope to all of the claims or theories outside the biological sciences that have been described as "Darwinian"), and chooses to clarify it by laying out what we might think of as a set of minimal conditions for a position to qualify as "Darwinian." For Gayon, then, "'Darwinian' is taken to mean any interpretation of evolution as being the product of the gradual modification of species, predominantly guided by a process of natural selection functioning on a field of intra-populational variation" (Gayon, 1998, p. 4). As we have seen, this definition certainly leads to the identification of a number of non-Darwinian theories in the authors we've discussed. If a theory must invoke gradual modification in order to count as Darwinian, then Galton's Darwinian stripes are in question, and Bateson is a clear anti-Darwinian. But deciding what exactly is to count as Darwinian using Gayon's criteria becomes difficult when applied to these historical actors, as a number of the terms that Gayon uses here (such as "a field of intra-populational variation") translate only with difficulty back into the language of biologists working in the late 19th or early 20th centuries.

A further conundrum, then, concerns whether or not it is reasonable to call such theories "Darwinism"—or whether Darwinism is, indeed, a useful theoretical category in the first place. The example of Galton, again, might be the most instructive. If one of Darwin's direct successors, someone with whom Darwin himself traded reams of correspondence, and who sincerely believed that he was continuing the Darwinian legacy, is ruled out as presenting a "non-Darwinian" theory of evolution, we would be within our rights to question whether the term in fact has any meaning left worth preserving. In that case, abandoning "Darwinism" (and the "eclipse of Darwinism" along with it) might be the most historically faithful choice. Today's theories of "Darwinian" evolution surely satisfy all of Gayon's criteria—he notes that he is unapologetically engaging in rational reconstruction, attempting to evaluate how it is that the history of science has led us to what we now recognize as our best extant scientific theory (Gayon, 1998, pp. 6–8)—but if such a category leads us to counter-intuitive conclusions about even central historical figures, then perhaps it is doing more harm than good in a narrative like mine.

Next, we should consider the very notion of an "eclipse" itself. The idea has been masterfully, and skeptically, analyzed by Mark Largent (2009). As he argues, the concept of an "eclipse" has powerful rhetorical force. It is no coincidence that it was first deployed in Julian Huxley's book that also coined the term "Modern Synthesis"—it serves as a way to drive a wedge between the efforts of Synthesis-era biologists and those who had preceded them. Work done during an eclipse is, perhaps definitionally, benighted. On such a metaphor, the lack of a clear understanding of genuine "Darwinism" led figures in this period to make fundamental mistakes, misread genetics, and advocate eugenics, all of which could now be cleanly and justifiably left behind by contemporary evolutionary science. Further, as Largent notes, the sources drawn on to demonstrate the existence of such an "eclipse" are often not particularly diverse—a number of authors point only to Vernon Kellogg's *Darwinism To-Day* (1907) as emblematic of the extent to which natural selection had been cast aside. But when we examine Kellogg's text, the story isn't quite so clear. Consider, for instance, his presentation in the introduction of the trend toward a "most careful re-examination or scrutiny of the theories connected with organic evolution" (Kellogg, 1907, p. 1):

> *...there are being developed and almost feverishly driven forward certain fascinating and fundamentally important new lines, employing new methods, of biological investigation. Conspicuous among these new kinds of work are the statistical or quantitative study of variations and that most alluring work variously called developmental mechanics, experimental morphology, or, most suitably of all because most comprehensively, experimental biology.*
>
> **(Kellogg, 1907, pp. 1–2)**

While this paragraph can certainly be read as carrying the implication that all such efforts have as their goal the rejection of Darwin's approach to natural selection, I think a more apt reading places them squarely within the textbook tradition that we saw in Chapter 5. Was Darwin's theory under scrutiny? Of course. There was no way that the re-discovery of Mendel and the attempt to integrate his insights into the process of evolution couldn't have led to that outcome. But these "feverish" efforts are congruent with a project to integrate those insights into a theory of evolution—assuredly expanded and adapted—which still owed much to Darwin, even if it might not be "Darwinian" in Gayon's sense.

In short, with Largent, I think that my narrative here offers yet more evidence for his claim that

> *the notion of an eclipse of Darwin or eclipse of Darwinism is seriously problematic. The idea is inappropriately deterministic. It obscures the work done by two generations of biological researchers. It is metaphor that implies a dark age of evolutionary thought, and it serves as the basis for the production of a discontinuous history of evolutionary thought. Current use of it is much like the term social Darwinism: popularized by the succeeding generation, employed to denigrate and build distance, and leading to a fundamental misinterpretation of key texts.*
>
> **(Largent, 2009, p. 15)**

All of which constitutes my first serious disagreement with Gayon: insofar as "the eclipse of Darwinism" is one of the structuring principles of his analysis, I think this predisposes him to see discontinuity where I've argued for continuity, and to look past what I think are some of the most exciting, incremental transformations of the period.

This brings us to Gayon's second major interpretive axis: the concept of heredity. For Gayon, the key to explaining the eclipse is this notion's shifting meaning. As we move from Darwin, who endeavored (pangenesis aside) to treat heredity's role in natural selection as a black box, to the biometrical conception of ancestral heredity, to the re-construction of selection in the Modern Synthesis on the basis of a newly described genetic foundation for heredity, it is this movement, Gayon claims, that is crucial to the understanding of natural selection. As I briefly discussed earlier, that concept has not been my focus here. In this case, I don't disagree with Gayon—I think that a history of this period centered on heredity is extremely important.[a] I intend for my interpretation to be largely complementary to his—an evaluation of what happens if we take seriously the idea that chance and probability have a more fundamental role in the construction and development of evolutionary thought.[b]

David Depew and Bruce Weber's *Darwinism Evolving: Systems Dynamics and the Genealogy of Natural Selection* (1995) also bears a clear affinity to my project.

---

[a] This is particularly so if one wants to understand the development and spread of Mendelian thought, an important pursuit that has explicitly been left out of my story. Crucial to this discussion in today's context will be a forthcoming book by Gregory Radick, *Disputed Inheritance*, which arrived to me in draft too late for me to be able to substantially engage with it in my work here.

[b] A smaller but still important point in comparison of my approach and Gayon's: while Gayon extensively discusses the work of the biometrical school, he stops his reading of Weldon's novel theoretical work in 1895 (Gayon, 1998, p. 224), and of Pearson's around 1898 (Gayon, 1998, p. 248), thereby missing the developments in late biometry that I have argued are vital for understanding their theoretical and philosophical approach.

For these two authors, precisely what marks out evolutionary theory is that "while successfully maintaining its autonomy from physics and other basic sciences, [Darwinism] has used explanatory models taken from the part of physics called dynamics to articulate, defend, and apply its core idea of natural selection" (Depew and Weber, 1995, p. xv). They are thus fellow travelers in the project to tell a history of natural selection on which commitments to mathematical structures—in their case, dynamical theories drawn from physics—are given much greater recognition in our understanding of the history of biology. In that sense, once again, I think my approach is largely complementary to theirs, an ambitious project to re-tell the history of biology as a progression of changes in the deployment of the kinds of explanatory resources upon which evolutionary theorists would draw to express the fundamental structure of their work.

The overlap, and lack thereof, between their understanding of dynamical models and my approach to concepts of chance and methods of statistics is, however, worth some exploration. First, they take the idea that Darwin is presenting a "Newtonian" model of natural selection even more seriously than I have in Chapter 1. On such a picture (of any science, evolution included), "whatever the entities are that conform to this model, they will have an inertial tendency of some sort driving them off on a tangent. This is diverted and shaped by an external force. The result is a system that maintains itself in equilibrium" (Depew and Weber, 1995, p. 9). Variation is the inertial tendency that drives organisms away from their clean divisions into species and genera, and sexual reproduction is the perturbing, mixing force that nonetheless holds them within that equilibrium. While one could quibble with the details, I think we are largely in agreement that Darwin produces a non-chancy, non-mathematized biology designed to appeal to readers of works like Lyell's *Principles of Geology* or Herschel's *Preliminary Discourse*.

Turning to the rest of the figures that I have analyzed here, however, one finds some discord. Depew and Weber borrow Hacking's contention that Galton stands out as the first to have given statistical arguments explanatory and predictive power on their own, not merely as descriptive summaries of properly causally effective underlying laws (Depew and Weber, 1995, p. 201). As I detailed in Chapter 2, this view is a much too restrictive take on Galton's work, and doesn't apprehend the extent to which Galton (and the biometricians after him) held fast to an unstable combination of independent statistical theorizing and a constant search for the foundations of that statistical theorizing in underlying causal accounts of individual inheritance. They quite rightly note, however, that Weldon's method encompasses both the statistical and the physiological, a process of "expected distribution, statistically significant deviation, followed by adaptationist explanation of the deviation based on field observation bolstered by experiment" (Depew and Weber, 1995, p. 214). But as we have also seen in William Provine and in Gayon's analyses, Depew and Weber map the primary contribution of the biometricians onto the debate between continuous and discontinuous variation as the source material for natural selection, and frame that debate in terms of a "war" between biometry and Mendelism, neither of which commitments I share.

The way in which they present Fisher's early works also merits serious consideration. Having underlined the relationship between explanatory resources in biology and those in physics, it is unsurprising that they focus on Fisher's background in statistical mechanics. As they write,

> *By the end of the nineteenth century, Newton's luminous explanation of the system of the world had been honorifically retired as an exemplar of great physics. Two new, but closely related, paradigm cases had taken its place: Maxwell's reduction of the phenomenological gas laws, relating temperature, pressure, and volume to statistically calculable collisions between millions of molecules, and, hard on its heels, Boltzmann's reduction of thermodynamics to more or less probable arrays of molecular motion.*

(Depew and Weber, 1995, p. 254)

Following on this broader trend in the sciences, they assert, Fisher makes two crucial moves. The first is one that I also discussed in Chapter 6; we might call it the replacement of an "actualist" ontology of extant organisms and populations, which had been so central for Pearson and Weldon as they offered accounts of natural selection, with a modal one of counterfactual populations in a space of evolutionary possibilities. In Depew and Weber's words, "when in the first half of the twentieth century the Darwinian tradition began to use dynamical models taken from statistical mechanics and thermodynamics, it changed its ontology, and expanded its problem-solving prowess by doing so" (Depew and Weber, 1995, p. 4). While I have already expressed some degree of doubt concerning the extent to which Fisher was influenced by his time studying statistical physics, I strongly agree with the general thesis at work here (echoed as well by Morrison, 2002)—that it is, at least in part, a shift in ontology that lets us understand the advantage of a Fisherian approach to natural selection over a Weldonian or a Pearsonian one.

But Depew and Weber then go further. Fisher's innovation is not just a change of ontology while leaving alone the rest of the theoretical superstructure. It is, more broadly, a result of re-fashioning a vision of biology de novo on the basis of the kinds of physical models at work in thermodynamics and statistical mechanics (and the concomitant concepts upon which those models rested, like new approaches to equilibrium and to energy). They write that, for Fisher, "Darwin's mature insistence on continuous variation is rendered consistent with particulate inheritance through the ability of the probabilistic concept of natural selection to yoke the two together" (Depew and Weber, 1995, p. 267). I agree entirely—I would only object that this realization was already present in the late work of Weldon, and was supported, if haphazardly and often in a technically unsophisticated manner, by many members of the biological establishment we surveyed in Chapter 5. Fisher's proofs that a consistent and effective version of natural selection can be derived on such a basis are entirely new and certainly impressive, and, as I have argued, in a real sense "complete" the biometrical program. But I am less certain that this marks a total rupture with tradition, in the sense of a discarding of previous explanatory models to be replaced with novel ones drawn from a new domain.

Looking outward

To conclude, I want to return to the major themes that I have described throughout the book, offering some hopeful steps toward generalizing them and indications in the direction of future work. The period from Darwin's first notebooks through to the Modern Synthesis (following on the failure of the eclipse metaphor, we perhaps need to give it a new moniker; Largent, 2009 playfully borrows "interphase") is one of the most interesting in the history of evolutionary theorizing, and even now we have only just begun to really understand its depth and complexity.

One cluster of historiographical issues raised here, briefly mentioned already, can be understood in the context of re-thinking the grand narrative which has dominated the study of evolutionary biology from around 1880 to 1910: the debate between the biometricians and the Mendelians. First and foremost, I hope to have demonstrated that our discussion has suffered for having placed too much focus on two particular, towering figures: Karl Pearson and William Bateson. As we saw in Chapters 3–5, the work of both was somewhat outside the broader current of biological thought in this period. More importantly, and more dangerously, each was significantly more polemical than many of their colleagues. It is certainly the case that for both of them, the conflict between biometry and Mendelism was the central pivot around which the rest of the life sciences turned. Neither would give an inch of ground to the other camp until their deaths. But we have been far too quick to assume that the entire field—the conceptual and analytic toolboxes of practicing biologists of all stripes—was thus subdivided and dominated in the same way. Even a figure like Weldon, who clearly enjoyed scrapping with enemies in public and private, was simultaneously filling research notebooks with cautious efforts at synthesis. A few authors, myself included, have attempted to present this argument in smaller, more isolated cases (Ankeny, 2000; Pence, forthcoming), and Yafeng Shan has also foregrounded this idea in a recent monograph (Shan, 2020). I hope that this turn—equal parts, I think, historiographic and philosophical—will allow us to approach this important and exciting period in the history of biology in a more comprehensive way.

There are three things that we might recover by pulling this body of work out from under Pearson and Bateson's shadows. First, we will find a much more robust sense of theoretical and practical continuity than the harsh rupture implied by a supposed war between the biometricians and the Mendelians, culminating in Fisher's synthesis. This is not an entirely novel observation, though it is assuredly recessive in the extant literature. Depew and Weber, for instance, note that "by the beginning of the 1920s" (i.e., well in advance of the publication of *GTNS*) we could already point to the existence of a genuine, independent discipline of population genetics (Depew and Weber, 1995, p. 243). John Turner has underlined the importance of several figures working from 1905 to 1909 in the particular case of mimicry which, as we have seen, was so important to Fisher (Turner, 1985, pp. 180–181). The same point has been made in a Kuhnian vein by Shan (2020). A re-interpretation of this historical record can yield a view of the development of evolution from 1830 to 1930 that looks much more like "normal science" (or, to make the inevitable pun, like "evolution") than it does like "revolution."

I think the story I have painted here, though, is even more robust than this. The hints at chancy, statistical theorizing present in Darwin were among those picked up by Galton. Galton's approach to evolution left a host of open questions which were precisely those pursued by Pearson and Weldon. The view of evolution which the two of them were cultivating, perhaps best exemplified by Weldon's late works, is extended, albeit piecemeal, by a whole variety of authors in the period after Weldon's death, and makes appearances throughout a number of often reprinted and widely read textbooks. All of this material is taken up by Fisher as he makes good on the promise of a statistical theory of the population-level action of natural selection on a Mendelian basis which the biometricians had been seeking a few decades prior. Of course, there are manifold technical and conceptual differences between these authors; the pages of this book have been filled with them. But a story which emphasizes those differences—or, worse, which emphasizes one particular difference (often either the Mendelian mechanism of heredity or the discontinuous nature of variation) to the exclusion of all others—threatens to obscure this important continuity. I am therefore left extremely doubtful that there was indeed a "Mendelian Revolution," at least within the parts of evolutionary theory that I have presented here—it seems far simpler to see the development of chance and statistics in evolution as a series of interlinked and related modifications to the theory that Darwin laid out in the *Origin*, by which it steadily transformed into the evolutionary biology of the Synthesis. Of course, a different history with a different focus might well come to a different conclusion—but I think it is profoundly important to see that there are at least some viable interpretations of evolutionary theory on which the Mendelian Revolution was largely taken in stride.

Second, we can recover a number of interesting "smaller" authors and trends within the literature of the day. Turner is exactly right when he argues that "minor authors often reflect the general current of muddled opinion among the majority of workers in a field better than the confident and incisive writings—right or wrong—of the great masters" (Turner, 1987, p. 329). Such an approach was also taken up in Kyung-Man Kim's trenchant study of the sociology of the biometry-Mendelism debate, which focused on the role of the conversion of key "paradigm articulators" to Mendelism (Kim, 1994), though this work, too, sacrifices too much of the interesting detail about these authors to the single question of whether or not they had successfully been converted to the Mendelian (read: "victorious") side (see, e.g., critiques of this "conversion" approach in Ankeny, 2000; Vicedo, 1995). As we saw particularly in Chapter 5 (though one can make an argument for the works of Weldon having been equally undervalued), the view of the landscape of evolutionary theory that one gets from the textbooks of authors like Lock, Goodrich, and Thomson looks quite different from what one would learn if one had stopped with Bateson's *Mendel's Principles*. Shining light on these figures is by no means to spend time in an intellectual backwater—rather, as we have seen here as well, this broader understanding of the field of play has helped us to unpack the relationship between Fisher and the extremely rich intellectual context in which he produced his early works. Such authors often have importance and impact that sheds light on traditional histories of "major" figures and episodes.

Third and finally, in taking seriously the various contributions that, together, constituted the development of chance in evolution, we have had the opportunity to unearth some of the radically different philosophical commitments—indeed, in several cases, philosophical innovations—to which these biologists turned as they constructed their evolutionary theories. I want to conclude by unpacking a few of these, though I should state at the outset that it is not my aim here to draw any morals for contemporary philosophical debates (I have done a bit of that work, as much of it as I'm comfortable to do, elsewhere; Pence, 2021). Rather, I believe that the approaches built in this period, by highly motivated, technically competent, and philosophically well-read scientists deeply invested in the justification of their work, constitute a broadly unrecognized and extremely exciting source of insight still relevant for philosophers today. This relevance derives not from some kind of direct Whiggish dialog between ancient voices and modern problems, but rather from the kinds of complex conceptual interactions that these figures believed must be explained in order to offer a justification for the utility and coherence of a statistical, a chancy, biology.

Galton's approach to chance in evolution positions him as a partisan of a Quetelet-inspired view that we might call "evolutionary statics." He was unable to see a way in which statistical dynamics might reasonably arise from a distribution of characters which seemed to have a preternatural ability to reproduce itself, unchanged, into the future. The only way out, he thought, consisted in adopting an uneasy if not contradictory blend of a statistical treatment of natural selection with a saltationism that argued that evolution often, if not always, proceeded by jumps, as Galton's "faceted stone" came to rest on a different, stable side. As the work of Provine (1971) perhaps most clearly demonstrates (though we saw the same idea expressed in both Gayon and in Depew and Weber), this question of saltationism can also ground another historical approach to the period—echoed, as we saw above, in Gayon's insistence that any scientific theory worthy of the name "Darwinism" must involve a gradualist understanding of natural selection. On this view, the real question concerning natural selection in play at this time is whether or not the variations that constitute the raw material upon which natural selection will act are small and continuous or large and discontinuous.

But two other things about Galton's treatment of evolution also stand out. First, Galton was instrumental—as we saw Gayon emphasize earlier (1998, pp. 106–115)—in making the very idea of heredity itself a part of the analysis of evolution. It was this move to "heredity" that allowed Galton, Gayon argues, to view the phenomena of inheritance, and specifically his particulate theory of latent and patent characters with its easy accommodation of reversion to distant ancestors and regression to the mean, in a new way. This, then, served as a crucial tool as Galton's successors attempted to understand the dynamics of these particulate characters, carried by something, they knew not what, in evolving populations. In the words of Depew and Weber (in the context of Fisher's theory), despite the extent to which the physical processes of inheritance were a black box throughout this period, all the authors in the Galtonian tradition knew that a statistical understanding of evolution would have to involve "whatever entities are most numerous and most independent" (Depew and

Weber, 1995, p. 514). Galton's view of heredity could at least offer a candidate for what that might look like, a candidate from which more fine-grained processes of evolutionary change could later be built.

Second, I think it is worth stopping to underscore the extent to which the problem of statistical statics and dynamics would have seemed to be an unprecedented challenge for Galton and his immediate successors. It is easy now, in a 21st century context, for us to comprehend dynamical theories that are grounded, at bottom, in statistical distributions; many of us were raised on science classes that at least gestured to how quantities in statistical physics like temperature or pressure operate this way (for all that such an explanation is almost never elaborated in the details; Sklar, 1992). Today, such an understanding is part and parcel of our conceptual furniture. But in an environment where precisely what was interesting about the early governmental uses of statistics, to Quetelet and others, was their reproducibility and predictability over time, it was not immediately apparent that this tool would be tailor-made for the description of long-term trends and changes within such a statistically described population. This, too, was an innovation—one, I have contended, developed in this context primarily by Weldon and Pearson—and one whose conceptual basis is not often studied or appreciated. Arguably, it was with Weldon's encounter with the relevant empirical data, first in the form of the potentially bifurcating populations of crabs in the bay at Naples and then, more importantly, as he compared the change in a single distribution over time in the Plymouth Sound crab data, that the possibility began to crystallize.

Several things prevented Weldon from making the connections himself (not least of which being his untimely death). Most significantly, Weldon's source data remained, until the very end of his career in his attempts to accommodate Mendelism, empirical measurements on real-world populations. Having not yet been able to make sense of the way in which a Galtonian, particulate system of inheritance could give rise to statistical distributions of continuous characters, he had no underlying framework which would have provided him with theoretical distributions of anything else—he lacked the ontology he needed to be able to theorize at the level of generality that Fisher could, for instance, on the basis of his assumption that phenotypic characters trace to large numbers of independently assorting Mendelian factors. For a general theory to be possible in a given domain, we must have at least seized on an underlying ontology that allows us to draw up the conceptual space in which the theory is taken to operate. In Galton, Pearson, and Weldon's use of actual individuals drawn from actual breeding populations, such an ontology simply wasn't to be found (while, on the contrary, the Mendelian approach of Bateson and others shines). In general, though, this process by which we moved from statistics as statics to statistics as dynamics is worthy of more attention from philosophers of science, and points to interesting relationships between statistical theorizing, state spaces and their connection to modality, and the world which those statistical theories are taken to describe. In some sense, it seems, it was not the failure to identify the right kinds of real-world objects to which a statistical selection explanation might apply; it seems rather to have been a failure to identify the right kind of ontology to let us generate relevant counterfactual cases. Fisher's approach, in the end, allowed us to think about the actual population as just one member of (the right kind of) larger set of possible populations, a breakthrough that just wasn't apparent to the early biometricians. Perhaps it is here that we can see most clearly the impact of Fisher's training in statistical physics.

Pearson's positivism has been extensively discussed elsewhere (e.g., Depew and Weber, 1995, pp. 211–213; Gayon, 1998, pp. 290–291; Pence, 2011; Sloan, 2000), and so I will not dwell upon it here, other than to note that ascribing that same positivism to Weldon, as Gayon does, is clearly mistaken. Weldon's mature philosophy of science, on the other hand, is much less well known. The approach that he took to statistical theorizing over populations of real-world, extant organisms entailed, as he himself thought, an eminently practical and pragmatic problem for biologists. For we didn't yet know enough about those populations to be able to abstract away from them in the right kinds of ways. Only when we could confidently delineate what within our measurements of the biological world is real variation and what results from experimental error would we be able to clarify and classify our experimental data in the way that the physical sciences were already able to do. As I noted in Chapter 4, it is not clear whether Weldon believed this was a permanent limitation or merely a temporary state of affairs in which biology found itself. At the same time as he argued that a deep and fundamental part of biological research was its reliance on partial and noisy data, he also seemed in his theoretical discussions to be moving in the direction of a perspective (via his work on chromosomal inheritance) that would be every bit as general as the physical theory that had grounded Lord Rayleigh's discovery of argon. Finding this underlying theory would be a way to help free biological thought from the constraints under which it labored.

A Weldonian biology, then, looks much like a hybrid between the type of abstract statistical work that would have been familiar to Pearson or Fisher and the sort of detailed search for underlying physiological causes that would have delighted even the most ardent Mendelian. The combination of these two elements—which, of course, Weldon himself never completely effected—typifies exactly the kinds of challenges with which contemporary biology still wrestles, and which give rise to a number of debates about the relationship between biological theory and biological causation that are common in the philosophy of biology today (see, e.g., Uller and Laland, 2019). They have been with us ever since the first efforts to derive a statistical approach to natural selection.

Such a theory would still not be reached in the dozen years after Weldon's death. There was wide recognition that biometry and Mendelism had each produced bodies of work extremely well-suited to what they did, presenting, on the one hand, nearly incontrovertible facts about the nature and structure of a variety of populations (along with demonstrating connections between them like correlation and regression), and, on the other, offering equally incontrovertible explanations of the way that, in many circumstances in both agricultural and natural populations, characters were passed from parents to offspring and expected results were obtained in small- to medium-sized breeding populations. Squaring the two, and even doing so in a way that made room for a statistical, population-level theory of natural selection, was a widely held theoretical objective, and a host of vital steps were made during this period toward obtaining it. In particular, Mendelism steadily came to take on precisely the ontological role in support of a generalized theory of evolution that had gone unfilled for the biometricians. As we saw in the last chapter, this would be a crucial innovation by Fisher, who would on this basis be able to describe the distributions of Mendelian factors within populations in precisely the sort of way that could finally ground a general approach to selection. We have, however, little independent philosophical work to consider during this period; normal science, indeed, is the order of the day.

Lastly, we have Fisher. Of course, it is by no means breaking news that R. A. Fisher's work is philosophically rich; it is equally no great shock that it is quite difficult to approach, and even in only exploring his early work through the publication of *GTNS*, we have had to reconcile a number of influences that are, to say the least, quite diverse. Having already mentioned above Fisher's key inspiration in positioning a version of Mendelism as the state space in which natural selection could operate, I should note the final significant step in the philosophical story: his move to seeing this state space not solely in terms of the actual combinations of Mendelian factors present at a given time, but also as including all possible such combinations. For now we have not only ontological considerations at the fore, but modal considerations as well. The Fundamental Theorem reconnects this state space to actual populations, but at the cost of defining natural selection in an exceedingly indirect manner: as the relationship between a portion of the population's overall growth rate and the amount of variability it contains, measured with respect to fitness. As the theory has become more general, and successively whittled natural selection away from the other influences which operate on every population, the connection between theory and world has become more tenuous. Little wonder, then, that contemporary evolutionary theory would leave plenty of work for today's philosophers of biology.

I do recognize that my approach here takes away a bit of the magic surrounding the development of chance and statistics in evolution. There is something appealing about the "eclipse" story, of course. We begin with the "pure" insights of Darwin's original presentation of natural selection—Leonard Darwin, in a letter to Fisher, argued that the first edition of the *Origin of Species* was superior precisely *"because it was written before my father had been subject to any criticism whatever"* (L.D. to R.A.F., late-September, 1926; Bennett, 1983, p. 81). That masterpiece is obscured for decades by a ghastly debate between Bateson and Pearson, upon which we may look back with a sense of enlightened regret, only to be revealed in its proper light once again as Darwin's unofficial grandson Fisher would, as a singular product of a troubled genius, reconcile Charles Darwin with the Abbot Mendel, deploying in the process a new, sophisticated account of statistics and chance drawn from theoretical physics. Of course, it is no surprise that the story here is less exciting than this fairy tale. I hope that the loss is more than compensated for by the complexity, depth, and insight of the real biological and philosophical work which took place over what is, I think, an undeniably compelling century of thought surrounding the nature and role of chance in evolutionary theory.

References

Ankeny, R., 2000. Marvelling at the marvel: the supposed conversion of A.D. Darbishire to Mendelism. J. Hist. Biol. 33, 315–347. https://doi.org/10.1023/A:1004750216919.

Bennett, J.H. (Ed.), 1983. Natural Selection, Heredity, and Eugenics: Including Selected Correspondence of R.A. Fisher With Leonard Darwin and Others. Clarendon Press, Oxford.

Bowler, P.J., 1992. The Eclipse of Darwinism: Anti-Darwinian Evolution Theories in the Decades Around 1900. Johns Hopkins University Press, Baltimore, MD.

Bowler, P.J., 2014. Francis Galton's saltationism and the ambiguities of selection. Stud. Hist. Phil. Biol. Biomed. Sci. 48B, 272–279. https://doi.org/10.1016/j.shpsc.2014.10.002.

Depew, D.J., Weber, B.H., 1995. Darwinism Evolving: Systems Dynamics and the Genealogy of Natural Selection. Bradford Books, Cambridge, MA.

Gayon, J., 1992. Darwin et l'après-Darwin: une histoire de l'hypothèse de sélection naturelle. Éditions Kimé, Paris.

Gayon, J., 1998. Darwinism's Struggle for Survival: Heredity and the Hypothesis of Natural Selection. Cambridge University Press, Cambridge.

Hodge, M.J.S., 1992. Biology and philosophy (including ideology): a study of Fisher and Wright. In: Sarkar, S. (Ed.), The Founders of Evolutionary Genetics. Kluwer Academic Publishers, Dordrecht, pp. 231–293.

Huxley, J.S., 1942. Evolution: The Modern Synthesis. Allen and Unwin, London.

Kellogg, V.L., 1907. Darwinism to-Day: A Discussion of Present-Day Scientific Criticism of the Darwinian Selection Theories, Together With a Brief Account of the Principal Other Proposed Auxiliary and Alternative Theories of Species-Forming. George Bell and Sons, London.

Kim, K.-M., 1994. Explaining Scientific Consensus: The Case of Mendelian Genetics. The Guilford Press, New York.

Largent, M.A., 2009. The so-called eclipse of Darwinism. In: Cain, J., Ruse, M. (Eds.), Descended From Darwin: Insights Into the History of Evolutionary Studies, 1900–1970. American Philosophical Society, Philadelphia, PA, pp. 3–21.

Monod, J., 1971. Chance and Necessity: An Essay on the Natural Philosophy of Modern Biology. Alfred A. Knopf, New York.

Morrison, M., 2002. Modelling populations: Pearson and Fisher on Mendelism and biometry. Br. J. Philos. Sci. 53, 39–68. https://doi.org/10.1093/bjps/53.1.39.

Pence, C.H., 2011. "Describing our whole experience": The statistical philosophies of W. F. R. Weldon and Karl Pearson. Stud. Hist. Philos. Biol. Biomed. Sci. 42, 475–485. https://doi.org/10.1016/j.shpsc.2011.07.011.

Pence, C.H., 2021. W.F.R. Weldon changes his mind. Eur. J. Philos. Sci. 11, 61. https://doi.org/10.1007/s13194-021-00384-3.

Pence, C.H., forthcoming. How not to fight about theory: the debate between biometry and Mendelism in *Nature*, 1890–1915. In: De Block, A., Ramsey, G. (Eds.), The Evolution of Science. University of Pittsburgh Press, Pittsburgh, PA.

Plutynski, A., 2009. The modern synthesis. In: Routledge Encyclopedia of Philosophy. Routledge.

Plutynski, A., Vernon, K.B., Matthews, L.J., Molter, D., 2016. Chance in the modern synthesis. In: Ramsey, G., Pence, C.H. (Eds.), Chance in Evolution. University of Chicago Press, Chicago, pp. 76–102.

Provine, W.B., 1971. The Origins of Theoretical Population Genetics. Princeton University Press, Princeton, NJ.

Shan, Y., 2020. Doing Integrated History and Philosophy of Science: A Case Study of the Origin of Genetics. Springer, Cham.

Sklar, L., 1992. Philosophy of Physics. Westview Press, Boulder, CO.

Sloan, P.R., 2000. Mach's phenomenalism and the British reception of Mendelism. C. R. Acad. Sci. III 323, 1069–1079. https://doi.org/10.1016/S0764-4469(00)01255-5.

Sloan, P.R., Fogel, B., 2011. Creating a Physical Biology: The Three-Man Paper and Early Molecular Biology. University of Chicago Press, Chicago.

Smocovitis, V.B., 1994a. Disciplining evolutionary biology: Ernst Mayr and the founding of the Society for the Study of Evolution and *Evolution* (1939–1950). Evolution 48, 1–8. https://doi.org/10.2307/2409996.

Smocovitis, V.B., 1994b. Organizing evolution: founding the society for the study of evolution (1939–1950). J. Hist. Biol. 27, 241–309. https://doi.org/10.1007/BF01062564.

Smocovitis, V.B., 1996. Unifying Biology: The Evolutionary Synthesis and Evolutionary Biology. Princeton University Press, Princeton, NJ.

Turner, J.R.G., 1985. Fisher's evolutionary faith and the challenge of mimicry. Oxf. Stud. Evol. Biol. 2, 159–219.

Turner, J.R.G., 1987. Random genetic drift, R. A. Fisher, and the Oxford School of ecological genetics. In: Krüger, L., Gigerenzer, G., Morgan, M.S. (Eds.), The Probabilistic Revolution, Volume 2: Ideas in the Sciences. Bradford Books, Cambridge, MA, pp. 313–354.

Uller, T., Laland, K.N. (Eds.), 2019. Evolutionary Causation: Biological and Philosophical Reflections. The MIT Press, Cambridge, MA.

Vicedo, M., 1995. What is that thing called Mendelian genetics? Soc. Stud. Sci. 25, 370–382.

Index

Note: Page numbers followed by *f* indicate figures.